Protein Crystallization under the Presence of an Electric Field

Protein Crystallization under the Presence of an Electric Field

Special Issue Editor

Abel Moreno

MDPI • Basel • Beijing • Wuhan • Barcelona • Belgrade

MDPI

Special Issue Editor
Abel Moreno
Universidad Nacional Autónoma de México
Mexico

Editorial Office
MDPI
St. Alban-Anlage 66
4052 Basel, Switzerland

This is a reprint of articles from the Special Issue published online in the open access journal *Crystals* (ISSN 2073-4352) from 2017 to 2018 (available at: https://www.mdpi.com/journal/crystals/special_issues/protein_crystallization).

For citation purposes, cite each article independently as indicated on the article page online and as indicated below:

LastName, A.A.; LastName, B.B.; LastName, C.C. Article Title. *Journal Name* **Year**, *Article Number*, Page Range.

ISBN 978-3-03897-519-9 (Pbk)
ISBN 978-3-03897-520-5 (PDF)

Contents

About the Special Issue Editor

Abel Moreno was awarded a B.Sc. in Chemistry from the Autonomous University of Puebla (Mexico) in 1990 and a Ph.D. in Chemistry from the University of Granada (Spain) in 1995. Nowadays, Dr. Moreno is full Professor of Biological and Physical Chemistry at the Institute of Chemistry of the National Autonomous University of Mexico (UNAM) in Mexico City. He has been distinguished as a member of the National System of Researchers of Mexico (SNI) at level 3 (the highest category of Mexican scientists), a member of the Mexican Academy of Sciences, the New York Academy of Sciences and a member of the Mexican and American Chemical Societies. Prof. Moreno has been a visiting professor at the University of Cambridge (United Kingdom, 2009) and at the University of Strasbourg (France, 2003–2004). Dr. Moreno has been a visiting scientist at the University of Luebeck and at the Institute of Crystal Growth (IKZ) Berlin (Germany, February 2004), at the University of Tohoku (Japan, Autumn 2003), at Imperial College London (United Kingdom in Summer–Autumn 1999 and 2000), and at the University of California Riverside (USA, 1997). Dr. Abel Moreno has published more than 97 papers in prestigious international journals. He is the author of 15 book chapters and 6 books on his specialties in biological crystallogenesis, crystallochemistry, and biomineralization processes. Prof. Moreno was the former President of the International Organization for the Biological Crystallization from September 2010 to September 2012 (IOBCr). He is also member of the international advisory board of the Commission of Crystal Growth and Characterization of Materials of the International Union of Crystallography. Prof. Moreno is currently the President of the Mexican Society of Crystallography from 2018–2021. He is also member of the Advisory board of the Latin America Asia Africa and Middle East Program (LAAAMP) of IUCr–UNESCO–IUPAP. Prof. Moreno is member of the Editorial Board of the journal Progress in Crystal Growth and Characterization of Materials, Editor for the Latin America section of the newsletter of the International Union of Crystallography and Editor-in-Chief of the section Biomolecular Crystals of the journal Crystals (MDPI, Switzerland).

Preface to "Protein Crystallization under the Presence of an Electric Field"

Today, the use of electrically-assisted protein crystallization methods using DC or AC have revealed that crystals grow on average better in terms of crystal quality and orientation to the cathode (when the protein molecule is positively charged), compared to the crystals grown on the anode (which is a negatively charged protein molecule). These electro-assisted crystallization techniques produce a remarkable influence on nucleation and allow us to grow protein crystals as a function of controlled physical (like temperature) and chemical parameters (like pH, concentration and chemical potential) under the influence of DC or AC electric fields. It has also permitted the isolation of polymorphs of proteins, as published elsewhere. According to recent publications, AC current could affect not only the number of crystals, but also their size, depending on its frequency. To date, the trend in crystal growth for model studied proteins is as follows: the higher the AC, the higher the number of crystals. This trend can be used in the near future for applications to free electron laser (XFEL) experiments to solve complicated protein structures using the fourth generation of synchrotrons all over the world. There has recently been an important number of publications focused on this topic, however, there is lack of studies of some important physical and chemical aspects that have not yet been published. For these reasons, this book "Protein Crystallization in the Presence of an Electric Field" from the journal *Crystals* (ISSN 2073-4352) covers these novel trends in protein nucleation control and the crystal growth of biological macromolecules.

Abel Moreno
Special Issue Editor

crystals

MDPI

Review

Recent Insights into the Crystallization Process; Protein Crystal Nucleation and Growth Peculiarities; Processes in the Presence of Electric Fields

Christo N. Nanev

Rostislaw Kaischew Institute of Physical Chemistry, Bulgarian Academy of Sciences, 1113 Sofia, Bulgaria; nanev@ipc.bas.bg; Tel.: +359-2-8566458

Academic Editor: Abel Moreno
Received: 22 September 2017; Accepted: 12 October 2017; Published: 15 October 2017

Abstract: Three-dimensional protein molecule structures are essential for acquiring a deeper insight of the human genome, and for developing novel protein-based pharmaceuticals. X-ray diffraction studies of such structures require well-diffracting protein crystals. A set of external physical factors may promote and direct protein crystallization so that crystals obtained are useful for X-ray studies. Application of electric fields aids control over protein crystal size and diffraction quality. Protein crystal nucleation and growth in the presence of electric fields are reviewed. A notion of mesoscopic level of impact on the protein crystallization exercised by an electric field is also considered.

Keywords: protein crystallization; classical and two-step nucleation mechanisms; impact of electric fields on the protein crystallization; external and internal electric fields; number density; size and quality of protein crystals

1. Introduction

Crystallization is a ubiquitous process occurring in nature, technology, and even in biology (e.g., bio-mineralization of bone, teeth, and shells). Crystals are present in both healthy (insulin) and ailing humans (formation of kidney and gall stones, uric acid crystals in gout, amyloid fibrils and insoluble plaques, the latter been considered the causative agents in in some neurodegenerative diseases, such as Alzheimer's disease and Parkinson's disease). Due to the vital importance of proteins for all living organisms and protein biochemical involvements [1], three-dimensional protein structures attract a continuously growing research interest. When it comes to understanding the mechanisms of life and human genome, as well as developing novel protein-based pharmaceuticals [2], these structures prove to be of essential importance.

The preferred techniques for determination of three-dimensional protein molecular structures involve X-ray (and neutron) diffraction. However, they require crystals that are large enough [3], and well-diffracting. The well-known difficulties encountered in attempting to grow crystals of newly expressed proteins have prompted exploration of many diverse crystallization approaches. Related to these approaches is also the study of external physical factors, such as magnetic and electric fields (EFs); proper crystallization conditions can be fine-tuned using variation of both direct current (*dc*) and alternating current (*ac*) EFs. Pioneered by Aubry's group [4,5] some 20 years ago, protein crystallization under EF attracts an increasing attention, becoming a mature scientific branch today. Major contributions in this research field have been made by the teams of Moreno [6–17], Koizumi [18–29], Veesler [30–33], etc. (the list is not exhaustive). Traditionally, most of the experimental studies were performed with hen-egg white lysozyme (HEWL). Three reviews [10,31,34], and a book chapter [17] have already been published.

This paper focuses on the extensive research done on protein crystal nucleation and growth in EF [4–47]. In addition, the effect of EFs on other substances, e.g., the simplest possible amino acid glycine [48,49] and sucrose crystallization [50,51], is considered. For clarity, the paper also presents some basic crystallization process peculiarities.

Recent experimental and theoretical advancements supporting better protein crystallization understanding are discussed by Giege [52]. Some novel super-resolution techniques enabling a profound insight into the mesoscopic and molecular level scenarios of protein crystal nucleation and growth are elaborated as well [53]; such techniques are compared below. The aim of this comparison is to provide clues to which one of them may be applied for studying mesoscopic level processes of protein crystallization in EFs.

Currently, Atomic Force Microscopy (AFM), Laser Confocal Microscopy enhanced by Differential Interference Contrast (LCM-DIM) technique and Michelson Interferometry (MI) are used to visualize molecular-level surface processes and crystal morphology, as well as to measure crystal growth rates. In contrast to AFM, LCM-DIM is a non-invasive optical method. With several case studies, Sleutel et al. [54] have demonstrated the strengths, limitations and weaknesses of these three techniques. AFM, LCM-DIM and MI have different vertical, lateral and time resolutions, and their complementary application is considered to be highly advisable. AFM at molecular-level resolution is the most suitable technique for observing slow nanoscale growth processes. LCM-DIM is a microscale observation technique with nanoscale vertical resolution, useful for slow to medium growth processes, while MI is best suited for fast growth processes requiring only micrometric lateral resolution. The data acquired by means of these techniques are very useful for developing reliable physical models relevant to the crystal growth issues at hand [55].

Finally, the recently developed time resolved liquid-cell transmission electron microscopy (LC-TEM) proves to be a very powerful tool to study protein crystal nucleation and growth. For instance, Yamazaki et al. [56] have used LC-TEM to understand the mechanisms underlying the early stages of protein crystallization (see Section 2.1).

2. Crystal Nucleation

Spontaneous crystallization, which is a common practice for protein crystal growth, starts with formation of the smallest possible stable crystalline particles under the actual conditions, coined nuclei. To evoke nucleation, the equilibrated system needs to be supersaturated, i.e., it is necessary to change system energetic status [57]. For instance, the solution is cooled to evoke crystallization (with normal solubility temperature dependence) and vice versa, it is heated to crystallize (with retrograde solubility); water is heated to boil, it is cooled to freeze, etc. The measure for the degree of energy-change is the imposed supersaturation, $\Delta\mu = k_B \cdot T \cdot \ln(c/c_e)$, which is the driving energy for a new-phase nucleation and further growth (where k_B is the Boltzmann constant, T is the absolute temperature; and for solution, c is the actual concentration and c_e is the equilibrium one).

Being the first crystallization stage, the nucleation process predetermines important features of the subsequent crystal growth, such as polymorph selection (which is an issue of great interest for the pharmaceutical industry, because the same molecule may or may not have a therapeutic effect depending on the crystal polymorph), number of nucleated crystals, crystal quality, and crystal size distribution. Although benefiting from 120 years of research on small molecule crystal nucleation, the process with proteins is far from being thoroughly understood. Governed by some physical laws found initially for the small molecule crystallization, protein crystal nucleation is extremely complex. This complexity arises from the subtle interplay between physics and biochemical idiosyncratic features of proteins. It is the large size of their molecules and their highly inhomogeneous, patchy surface (that is essential for protein biological role) that evoke specificity to the molecular-kinetic protein crystal nucleation mechanism [58].

Despite nucleation importance, direct observation of critical nuclei has proven elusive, even nowadays. The reason is the inherent impossibility to observe the molecular scale acts of crystal nucleation. Protein crystal nuclei make no exception. Although formed by huge protein molecules, being still nanosized particles, they remain invisible by optical microscopy. AFM is able merely to visualize elementary acts during protein and virus crystal growth [59,60], while the sizes of the protein crystal nuclei are determined by means of thermodynamic estimations. Using LCM-DIM, Sazaki's group studies the 2-D nucleation kinetics of lysozyme [61] and glucose isomerase crystals (under high pressure) [62]. However, the main difficulty in the experimental study of the crystal nucleation arises out of the fact that it is impossible to distinguish the critical nuclei in the whole assembly of under-critical, critical and super-critical molecular clusters; cluster composition changes dynamically due to the constant growth/decay of differently sized clusters; and critical nuclei are not labeled. That is why they are indistinguishable under mere observation (only growth of 2D clusters is visualized by the most powerful observation methods, such as AFM, LCM-DIM and LC-TEM).

2.1. Evoking Nucleation; Classical Nucleation Theory (CNT) vs. Multi-Step Nucleation Mechanism

All nucleation phenomena, whether they proceed homogeneously or involve foreign particles, surfaces, EFs, etc., require formation of an interface between the old (mother) phase and the newborn condensed phase. Gibbs [63] has pointed out the major thermodynamic aspect of the nucleation process, namely, the large barrier to phase transition associated with the energy cost for creating this interface. The classical nucleation theory (CNT) is fundamentally based on interphase fluctuations, needed to surmount this barrier.

Although CNT has provided a reasonable explanation of the fluctuation-based nucleation mechanism and the nucleation barrier origin, in some cases it has failed to predict correctly nucleation rates, with deviations being of many orders of magnitude. Debating this lack of adequacy, researchers have proposed multistep nucleation mechanisms, formulated initially as a two-step nucleation mechanism (TSN). Unlike most small molecules, proteins can take diverse aggregation pathways that make the outcome of crystallization assays quite unpredictable. Ten Wolde and Frenkel [64] have predicted theoretically the existence of amorphous nuclei precursors. It was shown that the latter exist to a significant extent even in (under-)saturated solutions [65]. Whitelam [66] presents a molecular model designed to study crystallization in the presence and absence of amorphous intermediates. Based on computer simulations, he suggests tuning the relative strengths of the specific and nonspecific interactions, thus enabling the study of the relative efficiencies of various pathways leading towards the final crystalline state. Using dynamic light scattering and optical microscopy (for measuring apparent induction time for the occurrence of the first crystal), Ferreira et al. [67] have suggested a new version of the multistep nucleation mechanism where concurrent aggregation pathways competing with crystal nucleation are considered. As confirmed by dynamic light scattering analysis, the nucleation of lysozyme crystals is preceded by an initial step of protein oligomerization and by the progressive formation of metastable clusters. Unfortunately, however, dynamic light scattering is unable to discern structured (crystalline) from amorphous (or liquid) clusters.

Cluster formation pathways are largely discussed in the multistep nucleation theories, however, being the core of CNT, the fluctuation-based nucleation mechanism is not denied. While preserving CNT basic concept (a fluctuation-based nucleation mechanism), TSN denies only the simultaneous densification and ordering during a single nucleation event. According to the initial TSN formulation [68], mesoscopic droplets enriched in protein appear in the protein solution. Being only densified, this intermediate phase preserves some similarity to the mother phase. Then, due to the reduced surface tension, the phase-transition energy barrier is lowered bellow the one needed for direct transition mother-phase-to-crystal (occurring via the CNT mechanism). Thus, crystal nucleation is greatly facilitated in the intermediate dense liquid. The second step in TSN is the formation of crystal nuclei inside the highly-concentrated regions. Evidently, TSN resembles Ostwald's rule of stages, which stipulates that a thermodynamically less-stable phase appears first, and then a polymorphic transition toward a stable phase occurs.

The existence of amorphous nuclei precursors has been confirmed experimentally by Vivares et al. [69], Sauter et al. [70], and further by Schubert et al. [71]. Sleutel and Van Driessche [72] have observed a non-classical nucleation for the 3D liquid-to-crystal transition of glucose isomerase; local increase in density and crystallinity do not occur simultaneously, but rather sequentially. They have demonstrated that at high concentrations (~100 mg/mL), glucose isomerase can form mesoscopic liquid-like aggregates (the molecules retain enough mobility), which are potential precursors of crystalline clusters. These aggregates are stable with respect to the parent liquid and metastable compared with the crystalline phase. In contrast, glucose isomerase 2D crystal nucleation proceeds classically [73]; and the authors have proven the existence of a critical crystal size. Sleutel et al. also observed that, in this case, the interior of all clusters is in the crystalline state and the cluster dynamics are determined by single molecular attachment and detachment events [73].

According to most recent observations, however, the initial formulation of the TSN needs some redaction. This has been concluded by Yamazaki et al. [56] (who conducted experiments with LC-TEM). The authors have established that mesoscopic clusters, similar to those previously assumed to consist of a dense liquid and serve as nucleation precursors, are not liquid but amorphous solid particles consisting of lysozyme molecules. Moreover, lysozyme crystals never form inside them. Instead, nucleation events of orthorhombic lysozyme crystals attached to a silicon nitride window or to an amorphous solid particle are observed frequently. Nucleation is initiated with spherical particles which transform into faceted orthorhombic crystals. Under the tested experimental conditions, simultaneous formation of two lysozyme crystal polymorphs is observed, i.e., thermodynamically more-stable orthorhombic crystals and less-stable tetragonal crystals; the former grew further, while the latter dissolved. Moreover, orthorhombic crystals are more stable than amorphous solid particles under the experimental conditions. These observations clearly indicate that the amorphous solid particles act merely as a heterogeneous substrate that enhances the nucleation event, proceeding according to CNT. (In other words, the assumption that protein crystal nucleates heterogeneously on foreign particles of biological origin [74] is confirmed by these experiments). All this marks a significant departure from the initial formulation of the TSN [75].

Finally, an assessment of the balance between entropy and enthalpy for solute association to crystals is required to consider process thermodynamics. When incorporated into the crystal lattice, molecules lose the possibility to move freely in the solution. This results in entropy loss (that is due to the constrained translational and rotational degrees of freedom of the molecules), and disfavors crystallization. On the contrary, the release of some water molecules, attached to the contacting patches when crystalline bonds are formed, boosts system's entropy. Trapping and rearrangement of water also affect crystallization thermodynamics.

The entropic restriction is more important for protein crystallization due to the large size and complex shape of these molecules. Using LCM-DIM, Sleutel et al. [54] have determined the precise temperature dependent solubility of tetragonal lysozyme and glucose isomerase crystals. On this basis, the authors have characterized the thermodynamics of crystallization. Applying van´t Hoff equation, they have calculated the standard free energy, enthalpy and entropy of protein crystallization. The conclusions are that the entropic effect is compensated by the larger enthalpy change, and that the crystallization process is exothermic [54].

Comparing enthalpic and entropic contributions to the free energy of pre-nucleation cluster formation in the $CaCO_3$ system, Kellermeier et al. [76] (Table 1 in [76]) have noticed the minor costs in enthalpy linked to cluster formation, and conclude that the pre-nucleation cluster formation is predominantly driven by entropy. The entropic driving force is associated with the return of water to bulk solution; a gain in translational and rotational degrees of freedom is arising after water release from hydrated disordered precursors and water molecules move back into the surrounding solution. This drives the assembly of remaining solute molecules into an ordered (i.e., crystalline) structure.

Moreover, the key role of water release suggests that pre-nucleation cluster formation may be a common phenomenon in aqueous solutions.

3. Crystal Growth

After crystals nucleate, they start growing immediately. In this second crystallization stage, the crystals grow until solution depletion reaches a level which corresponds to zero supersaturation with respect to the smallest crystal in the system; this point marks the beginning of the so-called Ostwald ripening [57].

Multistep crystal nucleation pathways involving liquid-like, amorphous or metastable crystal precursors challenge our current understanding of crystallization by putting the question: Can also some metastable crystalline precursors play a role during the crystal growth? Being predicted by theoretical works and observed experimentally at nucleation, there is only some evidence that metastable crystalline precursors can also be relevant to the growth of the crystals. With proteins Sleutel and Van Driessche [72] have shown a surface cleansing, triggered by mesoscopic clusters of protein molecules formed in bulk solution. Sedimenting on the crystal surface, and merging with it, the clusters form expanding mounds containing a considerable number (ranging from 2 to > 100) of monomolecular steps. The expanding mounds trigger a step cascade that causes the self-purifying effect. If the impurity content of the arriving clusters is lower than the impurity concentration in the mother liquor, the steps propagating on the crystal surface, lead to its cleansing. The latter is a result of acceleration in the step velocity (which is due to the lower impurity concentration), and thus reduction of terrace exposure time with respect to impinging impurity.

Quite recently, Jiang et al. [77] have shown that disordered nanoscopic precursors can also play an active role in the stage of growth of organic compound crystals. Using in situ AFM on the {110} facets of a preexisting crystalline Glu.H_2O surface, the authors have observed that prenucleation clusters are involved during growth of DL-glutamic acid crystals. This non-classical scenario of growth proceeds through attachment and transformation of 3D nanoscopic precursors units (larger than the monomeric constituents) which finally transform into crystalline 2D nuclei; the latter eventually build new molecular layers by further monomer incorporation. Moreover, under a direct observation, the 3D nanospecies act as an initial material depot for subsequent epitaxial growth. The preexisting crystalline surface plays a crucial role in decreasing the barrier to epitaxial growth *via* heterogeneous nucleation. (On the opposite, due to lattice mismatch, the silicon surface lacks sufficient structural similarity to promote formation of 2D nuclei.) These results have been confirmed using three independent methods, such as electrospray ionization mass spectrometry, analytical ultracentrifugation and in situ AFM on an inert silicon substrate.

Sleutel et al. [78] have found that incorporation of growth units to crystal surface steps occurs through surface diffusion. As opposed to direct incorporation from solution where all these events need to operate in a concerted way and (therefore) result in a large activation barrier, surface diffusion is a two-step process where the barriers for adsorption and incorporation into the step are separated. Unexpectedly, proteins can also grow by the 2D nucleation mechanism even at low supersaturation due to the lack of active spirals on the crystal face [54]. Transition from 2D-nucleation to kinetic roughening of glucose isomerase crystals with supersaturation increase has been observed directly using LCM-DIM [79]. In such studies, computer simulations [80] can be very useful for predicting crystal growth.

3.1. Crystallization in the Presence of Electric Fields (EFs)

3.1.1. EFs Affect Protein Crystal Number Density and Improve the Quality of the Crystals Grown

Initially, EFs have been applied externally (e.g. [4,5,39]); that is to say, no contact between the electrodes and the solution. Systematic studies on the effect of a static EF have been carried out first by the Aubry's group [4]. The authors have also focused on providing a theoretical explanation of EF distribution (potential difference) inside the crystallizing system. Due to the high solution conductivity, the electrostatic potential change penetrates only in a thin solution layer but not in the solution bulk. With the aid of a vapor diffusion method, the authors show that EF suppresses HEWL crystal nucleation while simultaneously improving the diffraction quality of crystals grown (as estimated from the rocking-curve measurements conducted). It is assumed that EF directs HEWL molecules to fall oriented on the crystal surface, thus contributing to the improved crystal quality. The authors have also observed growth of HEWL crystals at the droplet surface near the cathode. Keeping in mind that, under the experimental conditions set, HEWL molecules are positively charged, this result follows the basic law in electrochemistry. When the voltage is higher than 1000 V, the drop starts to move towards the cathode. Subsequently, Aubry and coworkers [5] have measured directly protein concentration changes appearing upon application of an external EF and lead to a concentration gradient between the electrodes, the highest HEWL-concentration being observed near the cathode. This increased local supersaturation explains why EF affects protein crystallization.

Using custom-made 2D glass cells to crystallize HEWL, and simultaneously control temperature and substantially reduce convection, Nanev and Penkova [39] have applied an external high voltage (1500 V/cm) static EF. To ensure uniform EF in a flat condenser, two silver plates charged negatively and positively are pressed to the upper and bottom glass cell windows; in different experiments, the cathode is placed on the bottom or on the top of the cell. Under such conditions, EF is at its maximum in the solution adhering to the glass surface and decays rapidly towards solution bulk. It has been confirmed that HEWL crystals grow predominantly on the cathode side of the glass cell. In EF these crystals grow to visible sizes in less than 2 h, most of them being oriented with their c-axis normal to the supporting glass plate. Figure 1 shows highly predominant c-axis orientation of HEWL crystals (despite EF uniformity, deviations in the percentage of c-axis oriented crystals are noted at different places of the glass support). However, this preferred crystal orientation is reported to occur only at temperatures below 5–7 °C, down to 0 °C, while missing at higher temperatures, 18 °C and above.

It is logical to assume that EF orients HEWL molecules during the stage of protein crystal nucleation, and thus, predestinates crystal orientation. Applied through the (insulating) glass plate, only a small fraction of the external high *dc*-voltage affects HEWL crystallization. Therefore, its effect is observed merely at low temperatures, while, at higher temperatures, it is the thermal motion that prevails and disturbs protein molecule ordering, and prevents acquiring the preferred c-axis crystal orientation.

(a) (b)

Figure 1. Preferred orientation of HEWL crystals on the cathode side. Imposed EF = 1500 V/cm, 20 mg/ml HEWL and 0.7 M NaCl. Crystal sizes: 35 μm–70 μm. (**a**) Dark field image, $t = 0$ °C. (**b**) Bright field image, $t = 5$ °C [39] (with permission from [39]).

In addition to HEWL, Penkova et al. [40] have performed experiments on EF-assisted crystallization of ferritin and apoferritin in a sitting drop setup. The uncovered air/solution interface introduces complexity to the phenomenon; depending on the field strength, there occurs a solution stirring at rates of up to 100 μm s^{-1}. At slow solution flow rates, nucleation of ferritin and apoferritin crystals is suppressed, while faster stirring enhanced crystal nucleation of both proteins.

As already mentioned, high-quality and relatively large protein crystals are needed for protein structural crystallography based on X-ray (and neutron) diffraction. The excellent potential of external *dc*-EFs for growing such crystals has been confirmed most recently [41]. *Via* X-ray diffraction analysis it is proven that glucose isomerase crystals grown (by the microbatch method at room temperature) in the presence of *dc*-EFs of 1, 2, 4, and 6 kV are of higher quality as compared with crystals grown in the absence of EFs. Light microscopy observations indicate a decrease in crystal nucleation rate and an increase in crystal size with the increase in voltage applied. This could be seen in Figure 2 (Figure 7 of Rubin et al. [41]).

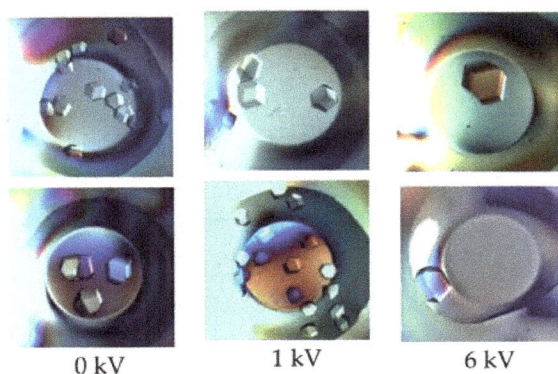

0 kV   1 kV   6 kV

Figure 2. Glucose isomerase crystals grown in the presence of different *dc*-EFs intensities for 48-h periods [41] (by permission of V. Stojanoff). The top and bottom panels are two typical experimental results.

An EF (electric potential difference) that injects a *dc* between electrodes immersed in the solution (conventionally used in the following consideration as "internal EF"), has been first applied in the laboratory of Moreno [6]. The authors merge capillary tubes and gels for studying electrochemically-assisted crystallization of lysozyme and thaumatin. Applying X-ray diffraction structure analysis for assessing crystallographic data of the grown crystals, Mirkin et al. [6] have revealed the good potential of this protein crystallization type. In addition, using the gel acupuncture method, cytochrome *c* crystallization has been expedited by a 15-day application of a constant current of 0.8 μA [9]. A review of the significant progress in the electrochemically-assisted protein crystallization achieved until 2008 has been presented by Frontana-Uribe and Moreno [10] (see Sections 4 and 5). The authors also point out the significant difference between the electrochemically-assisted protein crystallization and the true electrochemical reaction accompanying it. During the former, there is no redox reaction occurring between the protein and the inert electrodes (e.g., Pt and graphite) immersed in the crystallizing protein solution, but only an EF-steering of the proteins, similar to electrophoresis. This results in protein molecules concentrating near the electrode. In contrast, during an electro-crystallization per se, the current flows through the electrolyte because EF is applied to the electrodes immersed in a solution containing small-molecule ions. As a result, cations (positively charged ions) migrate toward the cathode and anions (negatively charged ions) move toward the anode. On the electrodes, the solution ions give up their charges, and the substance making up the ions is liberated; e.g., in water solutions, hydrogen and oxygen gases are released, respectively, on the cathode and the anode. Despite solution electrolysis however, no bubbles appear provided the current densities are low enough (because the small gas amounts dissolve in the solution).

dc-electrochemically-assisted batch crystallization of lysozyme and ferritin has continued to be the focus of Moreno's interest while working (as a visiting professor) with Sazaki in his lab in Japan [7]. They have observed that applying a *dc* of 2 μA flowing between electrodes of platinum wires, the number of deposited lysozyme crystals significantly decreases, while the size and the quality of crystals increase [7]. Nucleation induction time for crystallization also decreases. Apart from platinum, graphite electrodes have been used later in electrochemical Hull type cells adapted for protein crystallization [12]. Orthogonal cells of this type produce the largest size and the highest quality of lysozyme crystals in solution as well as in gel.

A novel transparent crystallization cell, composed of two indium tin oxide (ITO) covered (conductive) glass plates serving as electrodes, have been employed by Gil-Alvaradejo et al. [11]. X-ray diffraction analysis indicates an improved quality of lysozyme crystals grown at *dc* of 6 μA, and of ferritin crystals grown at *dc* in range of 2 and 6 μA. No conformational changes in the 3D protein molecule structures are noted. The strong adhesion of protein crystals enables their characterization by in situ AFM. Such ITO-covered-glass-electrodes have been widely used in subsequent studies of Moreno's group. For instance, the electrodes have been adapted to a sitting-drop vapor-diffusion crystallization setup applied for lysozyme and 2TEL-Lys crystallization [13]. As observed by the authors, the lysozyme crystals growing while attached to the cathode are larger than those grown in the absence of an electric current.

Similar transparent ITO-covered glass cells have also been used by Wakamatsu and Ohnishi to study HEWL crystallization [35]. The authors employ various voltage waves, including sine, triangular and step waves at frequencies of up to 15MHz, or *dc*. Wakamatsu [38] has tested ITO-based transparent cells for thaumatin crystallization in an extremely low internal *ac*-EF (a sine-wave voltage of 1.06 V at 20 Hz for 10 h). By means of the same transparent crystallization cell, a lysozyme molecule aggregate formation has been studied by applying an internal EF and a low-angle (<8°) dynamic light scattering technique [36]. The method and the apparatus used for characterization of the protein aggregation have been described in detail elsewhere [37]. Differently ITO-patterned on-glass-slides-electrodes have served as bottoms of micro fluidic devices with parallel electrodes prepared for the study of protein (lysozyme and insulin) crystallization [42].

In general, an EF-introduced potential energy landscape can be shaped by adapting different types of electrodes. To apply an *ac*-current injection field during protein crystallization, Hou and Chang [47] have constructed interdigitated and quadrupole Ti/Au electrodes to study the competition between gel and crystal formation at different voltages and frequencies. While applying EFs, there occurs a reduction in the nucleation site numbers, accompanied by a rapid increase in lysozyme crystal size. A method for protein crystallization involving the use of a microbatch under oil (where the crystallizing protein is contained in a small droplet of solution dispensed on electrically isolated electrodes) and a relatively low voltage (30–270 V) has been developed by Al-Haq et al. [46].

A crystallization cell with at least one sharp electrode has been proposed by Hammadi et al. [30]. To take control over the spatial and temporal location of the nucleation event, Hammadi et al. [32] have applied a localized (internal) *dc*-EF. The nanometer size of the electrode tip causes large EFs with steep field gradients and a high current density appearing in the solution close to the tip. Agarose gel is used to diminish convection. The experiments have been performed with bovine pancreatic trypsin inhibitor and lysozyme. To control the nucleation event location, the authors have employed the confinement effect of small droplets produced by microfluidics technologies coupled to external EF [33].

The combined effect of a 2 μA *dc*-EF and a 10 T homogeneous magnetic field has been studied as well [8]. A significant increase in both crystal size homogeneity and ratio of magnetically orientated crystals as compared with the controls is reported. Most recently, some sophisticated techniques for protein crystallization using *dc* (in the range of 2–6 μA) and *ac* (in the range of 2–8 Hz) have enabled control over crystal nucleation, size and orientation (the latter governed by a rather strong magnetic field of 16.5 T) [16]. These techniques allow for obtaining large and suitable crystals not only for high-resolution synchrotron X-ray, but also for neutron-diffraction crystallography. Combining adequately these modern experimental methods and involving nuclear magnetic resonance, the authors have determined the crystal lattice contacts for two model proteins, lysozyme and glucose isomerase. Another combination of 2 Hz *ac*-EF with a strong magnetic field of 16.5 T, and radiofrequency pulses of 0.43 µs, have been applied to study the growth of lysozyme crystals in both solutions and gels [15]. It has been established that, using all types of precipitants (salts, organic solvents, polyethylene-glycols, etc.), these crystallization conditions render high-quality lysozyme crystals. Thus, the authors have grounds to suggest this crystallization approach as an alternative of the agarose gel crystallization (although the polyethylene glycols are popular precipitating agent for obtaining high-quality crystals, agarose does not polymerize in the presence of polyethylene glycols).

In a series of works, Koizumi et al. [18–29] have achieved a breakthrough in EF-promotion of protein crystal nucleation. Previous experience and theoretical considerations has given them reason to realize that the difference in electrical permittivity of liquid and solid must be the factor that defines the degree of change (appearing due to the external EF) in the chemical potential, $\Delta\mu$ (nucleation driving energy). The larger the difference in the electrical permittivity, the greater the effect of an externally applied EF on the nucleation rate. Since this electrical permittivity difference depends on *ac*-EFs frequency, HEWL crystal nucleation rate increases significantly when a high frequency (1 MHz) external EF is applied, and decreases when an EF of 10 kHz is applied. The nucleation rate in a 500 kHz EF has been reported to be almost the same as the one in the absence of EF [18]. Thus, demonstrating both enhancement and retardation in the nucleation rate of HEWL crystals, the authors succeed in controlling effectively the process. Koizumi et al. [23] have also shown increase in crystal nucleation rate of porcine insulin by applying *ac*-EF of 3 MHz.

Furthermore, Koizumi et al. [19] have established that the electric double layer effect expressed at the solution drop/sealing oil interface is complementary to the effect of an external *ac*-EF. The authors have calculated that the external EF strength required to change the nucleation rate is $\sim 10^4$ V/cm, which is much larger than the experimentally applied EF of 800 V/cm. It turns out however that the important factor in changing the nucleation rate is the huge EF which is sustained in the electric double layer arising at the interface solution drop/oil; an electric potential difference of ~ 1 mV in a ~ 1 nm thick electric double layer corresponds to an external EF of $\sim 10^4$ V/cm (that is the calculated EF-strength). This finding has been experimentally confirmed in HEWL crystallization in an external *ac*-EF of 800 V/cm at frequencies of 1–9 MHz. The authors have observed that nucleation occurs only on the surface of the solution drop but not in solution bulk.

The greater the ionic strength of the solution (dependent on the precipitant used), the thinner the electric double layer at the solution drop/sealing oil interface, and the stronger the EF-effect exerted on the supersaturation (being a crystal nucleation driving factor) [20]. The effect of an external *ac*-EF on the HEWL crystal nucleation rate increases with the increase in the concentration of the precipitant used (in the case considered, the precipitant is $NiCl_2$). The reason is that the electric double layer becomes thinner with the increase in precipitant ionic strength [21].

Of prime importance to the growth of well-diffracting crystal is to understand how imperfections originate in protein crystals. By means of X-ray diffraction rocking-curve measurements of HEWL crystals grown in the presence/absence of an external 1 MHz *ac*-EF, Koizumi et al. [24,25] have shown that the quality of HEWL crystals improves when such *ac*-EF is applied during crystal growth; the local strain in the crystals decreases significantly. This could be explained by the electrostatic energy of EF that is added not only to the chemical potential but also to the entropy [81] of the liquid and solid phases. Lowering entropy of the solid, the external 1 MHz *ac*-EF imposes higher structural order in the growing crystals. The authors believe that certain disorder in the protein crystals may be attributed to water molecule irregularities included in the crystals. When such irregularities are eliminated, the external *ac*-EF improves the homogeneity of tetragonal HEWL crystals. In contrast, when applying an external *ac*-EF of 20 kHz, Koizumi et al. [27] have observed a local strain accumulation in tetragonal HEWL crystals grown during nine days. The theoretical analysis suggests that this frequency has a significant effect on the liquid entropy. The rocking-curve measurements of HEWL crystals grown in the presence/absence of an external *ac*-EF, carried out by Koizumi et al. [26,28], have shown that no dislocations but rather misorientation among subgrains in the crystals are responsible for their inhomogeneity and strain. The authors show that such misorientations result from incorporation of impurities (most probably lysozyme dimers) into protein crystals. These impurities are integrated into the steps on the crystal surface during crystal growth, and also between adjacent subgrains. To elucidate the effect of 1 MHz *ac*-EF on the elementary surface steps dynamics, Koizumi et al. [29] have measured (by means of microscopic observations) the growth rates of (110) and (101) faces of tetragonal HEWL crystals. The comparison of these data with the experimental results in the absence of EF shows a decrease in the growth rates of both faces under the applied *ac*-EF. This EF-effect is attributed to an increase in the effective surface energy of the step risers. The effects of a relatively low frequency *ac*-EF (less than 1 kHz) on HEWL crystal nucleation have been studied by Pan et al. [45], who conclude that such *ac*-EFs affect the mutual orientation between neighboring protein molecules.

3.1.2. Potential of EF in the Selection of Protein Crystal Polymorph

The dissolution rate of crystals and the delivery of drugs from crystal-containing formulations depend on the crystal polymorphic form. However, in a supersaturated solution, multiple polymorphs may appear. Hence, control over the polymorphic form is of importance for the pharmaceutical industry. Crystalline drugs must be manufactured in a specified polymorphic form only, and its purity is essential to its pharmaceutical application.

Evidently, the critical step in crystal polymorph selection lies in achieving control over nucleation stage. Having this in mind, Moreno's group [14] has chosen to apply a synergistic combination of complementary research approaches and tools (such as the use of temperature dependent solubility diagrams, X-ray diffraction and AFM, and molecular replacement method for structure determination) to study the temperature effect on glucose isomerase crystal polymorphism using a vapor diffusion setup with an ITO-electrodes crystallization cell. In doing so, the effect of 2 μA *dc*-EF on crystal nucleation and the optimal temperature for growing the best crystals has been explored. On the other hand, Kwokal and Roberts [44] have employed the electrochemical templating effect to look into the crystallization intricacies of different entacapone polymorphs using low magnitude *dc*-polarization of Au(100) and Au(111) interfaces.

Spherulitic crystals (crystals formed from thin needles growing radially outward from a center) may have a different morphology from the one of bulk crystals and, hence, dissolve differently. To study the number ratio between spherulitic and tetragonal HEWL crystals, growing together at very high supersaturations, Tomita et al. [22] have changed the frequency of the external 800 V cm^{-1} *ac*-EF to range from 1 to 3 MHz. The reason for their choice is the difference between electrical permittivity of crystal polymorphs that depends on the intra-crystalline water contained in the crystals. (The possible effect of Ostwald's rule of stages has remained beyond the scope of the study.)

3.2. EF Effects on Other Substances

EFs also affect crystallization of other substances. Preferred crystal nucleation of glycine γ polymorph has been observed by Aber et al. [48], who apply strong *dc*-EF to supersaturated aqueous glycine solution. The phenomenon is explained with the EF-induced orientation of the highly polar glycine molecules. Glycine polymorphism in the presence of a pulsed EF has also been explored by Di Profio et al. [49]. Their experimental results give grounds for the conclusion that the effect of different factors affecting glycine polymorphism can be graded as follows: EF > pH > supersaturation level > solute concentration.

Hu et al. [50] and Parniakov et al. [51] have looked into the effects of pulsed EF energy on sucrose crystal nucleation in supersaturated solutions. Parniakov and coworkers report a significant reduction in crystal nucleation induction time when pulsed EFs are applied as compared to the untreated sample. For a high voltage electric discharge-assisted nucleation, the authors observe a decrease in the effective crystallization time and an increase in the maximum rate of crystallization when EF strength and pulse number is increased. They conclude that the dependence of nucleation rate on EF strength, number of pulses being applied, supersaturation of sucrose solutions and solution flow rate provides an opportunity to control the size distribution and structure of sucrose crystals.

To be functionally active, the proteins require a minimum amount of water. Therefore, it is essential to explore EF effect on the surrounding water. Since the protein itself has an internal dipole moment, an EF may evoke conformational changes to its molecule. The coupling effect of protein surface charge and protein hydration water on protein overall dipolar response plays an important role. The electric force exerted on the protein dipole results in a torque that will rotate the protein and is likely to affect the diffusion motions of the hydration water. To study the dynamics of lysozyme and its hydration water, Favi et al. [43] have used quasielastic neutron scattering measurements in the presence/absence of external static EFs (D_2O and H_2O hydrated lysozyme is compared). The measurements reveal that the nano- to pico-second dynamics of the protein is unaffected by EF, possibly due to the stronger intra-molecular interactions compared to the maximum achieved field strength of 1 kV/mm. In addition, no appreciable quantitative enhancement of the diffusive dynamics of hydration water is observed. Finally, it is reported that EF of a solvated ion in water induces ordering in the surrounding water molecules, which, however, extends no further than several solvation shells [82].

As already mentioned, Koizumi et al. [27] have suggested that EF affects not only the protein molecules, but also the liquid entropy, probably because it impacts the water molecules. External EF makes them more polarized. Strong EF can also decrease hydrogen bond strength and even destruct such bonds in the water [83]. In other words, EF makes the water less structured. However, as noted above, the external EF is exponentially screened due to the high ionic strength of protein solutions that are under crystallization conditions. For the same reason, the electric double layer is relatively thin. Interestingly, static EF increases ice nucleation rate during crystallization of water. Theoretical calculations have shown that the application of such fields tends to decrease Gibbs free energy of the system, thus reducing the critical nucleus radius [84].

4. Perspective

Up to now, EF effects on growing protein crystals are studied by means of microscopic observations. The mesoscopic scale response of a crystallizing system to external EF can provide additional insights to the fine details of crystallization process under such conditions. A good candidate for this job seems to be the LCM-DIM, in combination with EF.

Enabling simultaneous in situ observation of single steps on a protein crystal surface and the crystal edges, LCM-DIM can render valuable information, distinguishing protein crystal growth from dissolution. Crystal dissolution starts from crystal edges, and the elementary surface steps propagate towards the center of the crystal face; and vice versa, growth is characterized by step propagation in reverse direction, towards the crystal edges. Therefore, growth and dissolution can be distinguished clearly by observing step propagation direction using LCM-DIM [54]. As is well-known, crystal surface growth occurs when supersaturation is imposed. With proteins having temperature dependent solubility, a crystal grows by setting solution temperature below (for normal solubility temperature dependence), respectively above (for retrograde solubility) the equilibrium temperature. Respectively, crystal dissolution occurs in the reverse cases, and the equilibrium temperature can be determined precisely; it corresponds to the case when the elementary surface steps do not move in either direction.

A quantitative rendition of the interrelation between EF effects and applied supersaturation is worth studying. Man obtains a decisive amount of information by seeing. Thus, an experimentum crucis would be imposing an EF on a preliminarily equilibrated (crystallizing) system, characterized by unmovable elementary surface steps. Switching on and off a *dc*-EF in a LCM-DIM observation will show whether EF has some impact on step dynamics or not. For instance, does EF energy [85] really change the equilibrium condition? The unambiguous answer to this question would be provided by a simple observation of the step behavior. If such steps start moving when applying a *dc*-EF, its impact will be obvious on a mesoscopic scale. If so, the sign and the amount of $\Delta\mu$ change that results from EF impact can be checked by observing whether the imposed EF prompts crystal growth or dissolution (respectively, whether EF accelerates or decelerates the movement of steps); and, if yes, to what extent? If such data are available, enthalpy and entropy of crystallization in presence of electric field can be calculated from the solubility, as described by Sleutel et al. [54]. It is of prime importance to check whether there is a change in this thermodynamic parameters that govern the crystallization process. (For fast crystal growth processes MI may be applied in a similar way.)

Likewise, EF effect (if any) on TSN mechanism of protein crystal nucleation and growth is worth exploring. It is to be hoped that the behavior of dynamic mesophases (amorphous metastable crystal precursors, according to TSN) under such conditions could be explained/predicted. In view of the importance of EF-impact, it is worth probing separately into protein crystal nucleation and growth. Detachment of the two consecutive stages, under simultaneous EF action, could be pursued. Extensive experimental observations as well as new ideas are needed to probe further into these issues.

Further research is required to clarify the effect that EF may exert on the protein crystallization by changing the hydration shells of protein molecules. At least a partial destruction of the letter is needed for crystal bond formation. It is the author's opinion, however, that the spatial distribution of water molecules in/around the protein lattice contacts depends on the kind of patches involved (hydrophilic or hydrophobic) and not so much on the EF. Computational approaches may help in elucidating the problem.

In addition, the complex response of different substances to EFs is worth studying to understand fully the mechanism of nucleation and crystal growth under such conditions [86].

5. Conclusions

Significant progress in studying the EF-assisted protein crystallization was achieved recently. The research in this field has been carried out using four main approaches, such as studies on the effects of external and internal EFs, by applying both *dc* and *ac*. Despite the progress, due to the enormous complexity of the process under such conditions, a detailed and uniform comprehension of EF-assisted protein crystal nucleation is still lacking. In particular, numerous mesoscopic and molecular scale crystallization mechanism details remain unknown. Needless to say, further upgrading knowledge with new information from experimental studies using sophisticated methods is required to support broader and profounder insights into this issue. Such information can be acquired from more precise measurements using novel super-resolution techniques, e.g., the LCM-DIM and MI. In addition, an appropriate combination of known approaches may improve our insight in the EF-effect on the crystallization process.

Understanding crystal nucleation and growth under EFs is important as far as practical application is concerned, e.g., use of *dc*-voltage during crystallization of food systems [84] and for inducing crystallization and to control crystal forms [87]. The application of *dc*-voltage during preservation of food systems (especially freezing) offers a new perspective to the food industry; during the freezing process, the *dc*-voltage promotes ice nucleation at a higher temperature and reduces the induction time. The freezing process under static EF produces smaller ice crystals in the food products, resulting in less freeze damage; it is thus expected to minimize cell disruption, to lessen the protein denaturation, and finally to preserve the texture of the fresh food after thawing [84].

Acknowledgments: The author would like to acknowledge the contribution of COST Action CM1402 Crystallize. This work is co-financed by the National Science Fund of the Bulgarian Ministry of Education and Science, under contract DCOST 01/22.

Conflicts of Interest: The author declares no conflict of interest. The founding sponsor had no role in the design of the study; in the collection, analyses, or interpretation of data; in the writing of the manuscript, and in the decision to publish the results.

References and Notes

1. Proteins enact a wide range of biological functions in the human body, such as catalysis, regulation, communication, mechanical support, movement, and transport.

2. For an excellent review on structure-guided drug discovery based on application of protein crystallography see "Blundell, T.L. Protein crystallography and drug discovery: Recollections of knowledge exchange between academia and industry. *IUCrJ* **2017**, *4*, 308–321.". The author would like to point out that his drive to explore topics such as crystalline insulin drug formulations, renin inhibitors, retroviral proteases (including AIDS antivirals), drug discovery in oncology and infectious disease, as well as for creating computer programs for generating reasonable protein structure models, has been significantly enhanced by the knowledge exchange between the pharmaceutical industry and academia.

3. Neutron crystallography requires growth of substantially larger protein crystals, greater than 0.1 mm^3 in size are preferred, i.e., 4–5 orders of magnitude larger than those used in synchrotron X-ray data collection. The power of neutron crystallography consists in higher precision by visualization of H-atoms (which play essential roles in macromolecular structure and catalysis), thus helping scientists to understand enzyme reaction mechanisms and hydrogen bonding.

4. Taleb, M.; Didierjean, C.; Jelsch, C.; Mangeot, J.P.; Capelle, B.; Aubry, A. Crystallization of proteins under an external electric field. *J. Cryst. Growth* **1999**, *200*, 575–582. [CrossRef]

5. Taleb, M.; Didierjean, C.; Jelsch, C.; Mangeot, J.P.; Aubry, A. Equilibrium kinetics of lysozyme crystallization under an external electric field. *J. Cryst. Growth* **2001**, *232*, 250–255. [CrossRef]

6. Mirkin, N.; Frontana-Uribe, B.A.; Rodriguez-Romero, A.; Hernandez-Santoyo, A.; Moreno, A. The influence of an internal electric field upon protein crystallization using the gel acupuncture method. *Acta Crystallogr.* **2003**, *D59*, 1533–1538. [CrossRef]

7. Moreno, A.; Sazaki, G. The use of a new ad hoc growth cell with parallel electrodes for the nucleation control of lysozyme. *J. Cryst. Growth* **2004**, *264*, 438–444. [CrossRef]

8. Sazaki, G.; Moreno, A.; Nakajima, K. Novel coupling effects of the magnetic and electric fields on protein crystallization. *J. Cryst. Growth* **2004**, *262*, 499–502. [CrossRef]

9. Mirkin, N.; Jaconcic, J.; Stojanoff, V.; Moreno, A. High resolution X-ray crystallographic structure of bovine heart cytochrome c and its application to the design of an electron transfer biosensor. *Proteins Struct. Funct. Bioinf.* **2008**, *70*, 83–92. [CrossRef] [PubMed]

10. Frontana-Uribe, B.A.; Moreno, A. On electrochemically assisted protein crystallization and related methods. *Cryst. Growth Des.* **2008**, *8*, 4194–4199. [CrossRef]

11. Gil-Alvaradejo, G.; Ruiz-Arellano, R.R.; Owen, C.; Rodríguez-Romero, A.; Rudiño-Piñera, E.; Antwi, M.K.; Stojanoff, V.; Moreno, A. Novel protein crystal growth electrochemical cell for applications in X-ray diffraction and atomic force microscopy. *Cryst. Growth Des.* **2011**, *11*, 3917–3922. [CrossRef]

12. Espinoza-Montero, P.J.; Moreno-Narváez, M.E.; Frontana-Uribe, B.A.; Stojanoff, V.; Moreno, A. Investigations on the use of graphite electrodes using a hull-type growth cell for electrochemically assisted protein crystallization. *Cryst. Growth Des.* **2013**, *13*, 590–598. [CrossRef] [PubMed]

13. Flores-Hernandez, E.; Stojanoff, V.; Arreguin-Espinosa, R.; Moreno, A.; Sanchez-Puig, N. An electrically assisted device for protein crystallization in a vapor-diffusion setup. *J. Appl. Crystallogr.* **2013**, *46*, 832–834. [CrossRef] [PubMed]

14. Martínez-Caballero, S.; Cuéllar-Cruz, M.; Demitri, N.; Polentarutti, M.; Rodriguez-Romero, A.; Moreno, A. Glucose isomerase polymorphs obtained using an ad hoc protein crystallization temperature device and a growth cell applying an electric field. *Cryst. Growth Des.* **2016**, *16*, 1679–1686. [CrossRef]

15. Rodríguez-Romero, A.; Esturau-Escofet, N.; Pareja-Rivera, C.; Moreno, A. Crystal Growth of High-Quality Protein Crystals under the Presence of an Alternant Electric Field in Pulse-Wave Mode, and a Strong Magnetic Field with Radio Frequency Pulses Characterized by X-ray Diffraction. *Crystals* **2017**, *7*, 179. [CrossRef]

16. Pareja-Rivera, C.; Cuéllar-Cruz, M.; Esturau-Escofet, N.; Demitri, N.; Polentarutti, M.; Stojanoff, V.; Moreno, A. Recent Advances in the Understanding of the Influence of Electric and Magnetic Fields on Protein Crystal Growth. *Cryst. Growth Des.* **2017**, *17*, 135–145. [CrossRef]

17. Moreno, A. Protein Crystallography, Methods and protocols. In *Advanced Methods of Protein Crystallization*; Wlodawer, A., Dauter, Z., Jaskolski, M., Eds.; Springer Protocols: Berlin, Germany, 2017; Chapter 3; pp. 51–76.

18. Koizumi, H.; Fujiwara, K.; Uda, S. Control of nucleation rate for tetragonal hen-egg white lysozyme crystals by application of an electric field with variable frequencies. *Cryst. Growth Des.* **2009**, *9*, 2420–2424. [CrossRef]

19. Koizumi, H.; Fujiwara, K.; Uda, S. Role of the electric double layer in controlling the nucleation rate for tetragonal hen egg white lysozyme crystals by application of an external electric field. *Cryst. Growth Des.* **2010**, *1*, 2591–2595. [CrossRef]

20. Koizumi, H.; Uda, S.; Fujiwara, K.; Nozawa, J. Effect of various precipitants on the nucleation rate of tetragonal hen egg-white lysozyme crystals in an AC external electric field. *J. Cryst. Growth* **2010**, *312*, 3503–3508. [CrossRef]

21. Koizumi, H.; Uda, S.; Fujiwara, K.; Nozawa, J. Control of effect on the nucleation rate for hen egg white lysozyme crystals under application of an external ac electric field. *Langmuir* **2011**, *27*, 8333–8338. [CrossRef] [PubMed]

22. Tomita, Y.; Koizumi, H.; Uda, S.; Fujiwara, K.; Nozawa, J. Control of Gibbs free energy relationship between hen egg white lysozyme polymorphs under application of an external alternating current electric field. *J. Appl. Cryst.* **2012**, *45*, 207–212. [CrossRef]

23. Koizumi, H.; Tomita, Y.; Uda, S.; Fujiwara, K.; Nozawa, J. Nucleation rate enhancement of porcine insulin by application of an external AC electric field. *J. Cryst. Growth* **2012**, *352*, 155–157. [CrossRef]

24. Koizumi, H.; Uda, S.; Fujiwara, K.; Tachibana, M.; Kojima, K.; Nozawa, J. Improvement of crystal quality for tetragonal hen egg white lysozyme crystals under application of an external alternating current electric field. *J. Appl. Cryst.* **2013**, *46*, 25–29. [CrossRef]

25. Koizumi, H.; Uda, S.; Fujiwara, K.; Tachibana, M.; Kojima, K.; Nozawa, J. Enhancement of crystal homogeneity of protein crystals under application of an external alternating current electric field. *AIP Conf. Proc.* **2014**, *1618*, 265–268.

26. Koizumi, H.; Uda, S.; Fujiwara, K.; Tachibana, M.; Kojima, K.; Nozawa, J. Control of subgrain formation in protein crystals by the application of an external electric field. *Cryst. Growth Des.* **2014**, *14*, 5662–5667. [CrossRef]

27. Koizumi, H.; Uda, S.; Fujiwara, K.; Tachibana, M.; Kojima, K.; Nozawa, J. Crystallization of high-quality protein crystals using an external electric field. *J. Appl. Cryst.* **2015**, *48*, 1507–1513. [CrossRef]

28. Koizumi, H.; Uda, S.; Fujiwara, K.; Tachibana, M.; Kojima, K.; Nozawa, J. Technique for high-quality protein crystal growth by control of subgrain formation under an external electric field. *Crystals* **2016**, *6*, 95. [CrossRef]

29. Koizumi, H.; Uda, S.; Fujiwara, K.; Okada, J.; Nozawa, J. Effect of an External Electric Field on the Kinetics of Dislocation-Free Growth of Tetragonal Hen Egg White Lysozyme Crystals. *Crystals* **2017**, *7*, 170. [CrossRef]

30. Hammadi, Z.; Astier, J.; Morin, R.; Veesler, S. Protein crystallization induced by a localized voltage. *Cryst. Growth Des.* **2007**, *7*, 1472–1475. [CrossRef]

31. Hammadi, Z.; Veesler, S. New approaches on crystallization under electric fields. *Prog. Biophys. Mol. Biol.* **2009**, *101*, 38–44. [CrossRef] [PubMed]

32. Hammadi, Z.; Astier, J.P.; Morin, R.; Veesler, S. Spatial and temporal control of nucleation by localized DC electric field. *Cryst. Growth Des.* **2009**, *9*, 3346–3347. [CrossRef]

33. Hammadi, Z.; Grossier, R.; Zhang, S.; Ikni, A.; Candoni, N.; Morin, R.; Veesler, S. Localizing and Inducing Primary Nucleation. *Faraday Discuss.* **2015**, *179*, 489–501. [CrossRef] [PubMed]

34. Al-Haq, M.; Lebrasseur, E.; Tsuchiya, H.; Torii, T. Protein crystallization under an electric field. *Crystallogr. Rev.* **2007**, *13*, 29–64. [CrossRef]

35. Wakamatsu, T.; Ohnishi, Y. Transparent cell for protein crystallization under low applied voltage. *Jpn. J. Appl. Phys.* **2011**, *50*, 48003. [CrossRef]

36. Wakamatsu, T.; Toyoshima, S.; Shimizu, H. Observation of electric-field induced aggregation in crystallizing protein solutions by forward light scattering. *Appl. Phys. Lett.* **2011**, *99*, 153701. [CrossRef]

37. Wakamatsu, T. Method and apparatus for characterization of electric field-induced aggregation in pre-crystalline protein solutions. *Rev. Sci. Instrum.* **2015**, *86*, 15112. [CrossRef] [PubMed]

38. Wakamatsu, T. Low Applied Voltage Effects on Thaumatin Protein Crystallization. *Trans. Mater. Res. Soc. Jpn.* **2016**, *41*, 13–15. [CrossRef]

39. Nanev, C.N.; Penkova, A. Nucleation of lysozyme crystals under external electric and ultrasonic fields. *J. Cryst. Growth* **2001**, *232*, 285–293. [CrossRef]

40. Penkova, A.; Gliko, O.; Dimitrov, I.; Hodjaoglu, F.; Nanev, C.; Vekilov, P. Enhancement and suppression of protein crystal nucleation due to electrically driven convection. *J. Cryst. Growth* **2005**, *275*, 1527–1532. [CrossRef]

41. Rubin, E.; Owen, C.; Stojanoff, V. Crystallization under an external electric field: A case study of glucose isomerase. *Crystals* **2017**, *7*, 206. [CrossRef]

42. Li, F.; Lakerveld, R. The influence of alternating electric fields on protein crystallization in microfluidic devices with patterned electrodes in a parallel-plate configuration. *Cryst. Growth Des.* **2017**, *17*, 3062–3070. [CrossRef]

43. Favi, P.M.; Zhang, Q.; O'Neill, H.; Mamontov, E.; Diallo, S.O. Dynamics of lysozyme and its hydration water under an electric field. *J. Biol. Phys.* **2014**, *40*, 167–178. [CrossRef] [PubMed]

44. Kwokal, A.; Roberts, K.J. Direction of the polymorphic form of entacapone using an electrochemical tuneable surface template. *CrystEngComm* **2014**, *16*, 3487–3493. [CrossRef]

45. Pan, W.; Xu, H.; Zhang, R.; Xu, J.; Tsukamoto, K.; Han, J.; Li, A. The influence of low frequency of external electric field on nucleation enhancement of hen egg-white lysozyme (HEWL). *J. Cryst. Growth* **2015**, *428*, 35–39. [CrossRef]

46. Al-Haq, M.; Lebrasseur, E.; Choi, W.; Tsuchiya, H.; Torii, T.; Yamazaki, H.; Shinohara, E. An apparatus for electric-field-induced protein crystallization. *J. Appl. Crystallogr.* **2007**, *40*, 199–201. [CrossRef]

47. Hou, D.; Chang, H.C. Ac field enhanced protein crystallization. *Appl. Phys. Lett.* **2008**, *92*, 223902. [CrossRef]

48. Aber, J.E.; Arnold, S.; Ward, M.D.; Garetz, B.A.; Myerson, A.S. Strong dc electric field applied to supersaturated aqueous glycine solution induces nucleation of the γ polymorph. *Phys. Rev. Lett.* **2005**, *94*, 145503. [CrossRef] [PubMed]

49. Di Profio, G.; Reijonen, M.T.; Caliandro, R.; Guagliardi, A.; Curcio, E.; Drioli, E. Insights into the polymorphism of glycine: Membrane crystallization in an electric field. *Phys. Chem. Chem. Phys.* **2013**, *15*, 9271–9280. [CrossRef] [PubMed]

50. Hu, B.; Huang, K.; Zhang, P.; Zhong, X.Z. Pulsed electric field effects on sucrose nucleation at low supersaturation. *Sugar Tech* **2015**, *17*, 77–84. [CrossRef]

51. Parniakov, O.; Adda, P.; Bals, O.; Lebovka, N.; Vorobiev, E. Effects of pulsed electric energy on sucrose nucleation in supersaturated solutions. *J. Food Eng.* **2017**, *199*, 19–26. [CrossRef]

52. Giege, R. What macromolecular crystallogenesis tells us—What is needed in the future. *IUCrJ* **2017**, *4*, 340–349. [CrossRef] [PubMed]

53. These techniques are applied because the resolution of the light microscope is limited to 200–300 nm (Abbe theory). Therefore, mesoscopic and molecular scale protein crystallization peculiarities are invisible by the conventional microscopy.

54. Sleutel, M.; Maes, D.; Van Driessche, A. Kinetics and Thermodynamics of Multistep Nucleation and Self-Assembly in Nanoscale Materials. In *Advances in Chemical Physics*; Nicolis, G., Maes, D., Eds.; Wiley-Blackwell: Malden, MA, USA, 2012; Volume 151, pp. 223–276.

55. It should be kept in mind, however, that supersaturations used to produce protein crystals suitable for X-ray diffraction are significantly higher than supersaturations that are optimal for AFM and LCM-DIM.

56. Yamazaki, T.; Kimura, Y.; Vekilov, P.G.; Furukawa, E.; Shirai, M.; Matsumoto, H.; Van Driessche, A.E.S.; Tsukamoto, K. Two types of amorphous protein particles facilitate crystal nucleation. *Proc. Natl. Acad. Sci. USA* **2017**, *114*, 2154–2159. [CrossRef] [PubMed]

57. Nanev, C.N. On some aspects of crystallization process energetics, logistic new phase nucleation kinetics, crystal size distribution and Ostwald ripening. *J. Appl. Cryst.* **2017**, *50*, 1021–1027. [CrossRef]

58. Nanev, C.N. Phenomenological consideration of protein crystal nucleation; the physics and biochemistry behind the phenomenon. *Crystals* **2017**, *7*, 193. [CrossRef]

59. Malkin, A.J.; Kuznetsov, Y.G.; McPherson, A. In Situ atomic force microscopy studies of surface morphology, growth kinetics, defect structure and dissolution in macromolecular crystallization. *J. Cryst. Growth* **1999**, *196*, 471–488. [CrossRef]

60. McPherson, A.; Kuznetsov, Y.G. Mechanisms, kinetics, impurities and defects: Consequences in macromolecular crystallization. *Acta Crystallogr. Sect. F Struct. Biol.* **2014**, *70*, 384–403. [CrossRef] [PubMed]

61. Van Driessche, A.E.S.; Sazaki, G.; Otálora, F.; Gonza'lez-Rico, F.M.; Dold, P.; Tsukamoto, K.; Nakajima, K. Direct and noninvasive observation of two-dimensional nucleation behavior of protein crystals by advanced optical microscopy. *Cryst. Growth Des.* **2007**, *7*, 1980–1987. [CrossRef]

62. Suzuki, Y.; Sazaki, G.; Matsumoto, M.; Nagasawa, M.; Nakajima, K.; Tamur, K. First direct observation of elementary steps on the surfaces of glucose isomerase crystals under high pressure. *Cryst. Growth Des.* **2009**, *9*, 4289–4295. [CrossRef]

63. Gibbs, J.W. On the Equilibrium of Heterogeneous Substances. *Trans. Connect. Acad.* **1879**, *3*, 108–248, 343–524. [CrossRef]

64. Ten Wolde, P.R.; Frenkel, D. Enhancement of protein crystal nucleation by critical density fluctuations. *Science* **1997**, *277*, 1975–1978. [CrossRef] [PubMed]

65. Gebauer, D.; Kellermeier, M.; Gale, J.D.; Bergström, L.; Cölfen, H. Pre-nucleation clusters as solute precursors in crystallization. *Chem. Soc. Rev.* **2014**, *43*, 2348–2371. [CrossRef] [PubMed]

66. Whitelam, S. Control of Pathways and Yields of Protein Crystallization through the Interplay of Nonspecific and Specific Attractions. *Phys. Rev. Lett.* **2010**, *105*, 88102. [CrossRef] [PubMed]

67. Ferreira, C.; Barbosa, S.; Taboada, P.; Rocha, F.A.; Damas, A.M.; Martins, P.M. The nucleation of protein crystals as a race against time with on-and off-pathways. *J. Appl. Cryst.* **2017**, *50*, 1056–1065. [CrossRef]

68. Vekilov, P.G. Nucleation of protein crystals. *Prog. Cryst. Growth Charact. Mater.* **2016**, *62*, 136–154. [CrossRef]

69. Vivares, D.; Kaler, E.; Lenhoff, A. Quantitative imaging by confocal scanning fluorescence microscopy of protein crystallization via liquid-liquid phase separation. *Acta Crystallogr. D Biol. Crystallogr.* **2005**, *61*, 819–825. [CrossRef] [PubMed]

70. Sauter, A.; Roosen-Runge, F.; Zhang, F.; Lotze, G.; Jacobs, R.M.J.; Schreiber, F. Real-time observation of nonclassical protein crystallization kinetics. *J. Am. Chem. Soc.* **2015**, *137*, 1485–1491. [CrossRef] [PubMed]

71. Schubert, R.; Meyer, A.; Baitan, D.; Dierks, K.; Perbandt, M.; Betzel, C. Real-time observation of protein dense liquid cluster evolution during nucleation in protein crystallization. *Cryst. Growth Des.* **2017**, *17*, 954–958. [CrossRef]

72. Sleutel, M.; Van Driessche, A.E.S. Role of clusters in nonclassical nucleation and growth of protein crystals. *Proc. Natl. Acad. Sci. USA* **2014**, *111*, E546–E553. [CrossRef] [PubMed]

73. Sleutel, M.; Lutsko, J.; Van Driessche, A.E.S.; Duran-Olivencia, M.A.; Maes, D. Observing classical nucleation theory at work by monitoring phase transitions with molecular precision. *Nat. Commun.* **2014**, *5*, 5598. [CrossRef] [PubMed]

74. Nanev, C.N. Kinetics and Intimate Mechanism of Protein Crystal Nucleation. *Prog. Cryst. Growth Charact. Mater.* **2013**, *59*, 133–169. [CrossRef]

75. This withstanding, cryo-TEM images also have shown that there is a second type of non-crystalline particles assembling lysozyme, which appear on amorphous solid particle or a container wall (thus, highlighting the role of heterogeneous nucleation), and that lysozyme crystals appear within such particles.

76. Kellermeier, M.; Raiteri, P.; Berg, J.; Kempter, A.; Gale, J.; Gebauer, D. Entropy drives calcium carbonate ion association. *Chemphyschem* **2016**, *17*, 3535–3541. [CrossRef] [PubMed]

77. Jiang, Y.; Kellermeier, M.; Gebaue, D.; Lu, Z.; Rosenberg, R.; Moise, A.; Przybylski, M.; Cölfen, H. Growth of organic crystals via attachment and transformation of nanoscopic precursors. *Nat. Commun.* **2017**, *8*, 15933. [CrossRef] [PubMed]

78. Sleutel, M.; Vanhee, C.; Van de Weerdt, C.; Decanniere, K.; Maes, D.; Wyns, L.; Willaert, R. The Role of Surface Diffusion in the Growth Mechanism of Triosephosphate Isomerase Crystals. *Cryst. Growth Des.* **2008**, *8*, 1173–1180. [CrossRef]

79. Sleutel, M.; Maes, D.; Wyns, L.; Willaert, R. Kinetic roughening of glucose isomerase crystals. *Cryst. Growth Des.* **2008**, *8*, 4409–4414. [CrossRef]

80. Anderson, M.W.; Gebbie-Rayet, J.T.; Hill, A.R.; Farida, N.; Attfield, M.P.; Cubillas, P.; Blatov, V.A.; Proserpio, D.M.; Akporiaye, D.; Arstad, B.; et al. Predicting crystal growth via a unified kinetic three-dimensional partition model. *Nature* **2017**, *544*, 456–459. [CrossRef] [PubMed]

81. For an extremely lucid and contemporary explanation of what entropy is see " Nishinaga, T. Thermodynamics -for understanding crystal growth-. *Prog. Cryst. Growth Charact. Mater.* **2016**, *62*, 43–57.".

82. Wilkins, D.M.; Manolopoulos, D.E.; Roke, S.; Ceriotti, M. Mean-field theory of water-water correlations in electrolyte solutions. *J. Chem. Phys.* **2017**, *146*, 181103. [CrossRef]

83. Shevkunov, S.V.; Vegiri, A. Electric field induced transitions in water clusters. *J. Mol. Struct. THEOCHEM* **2002**, *593*, 19–32. [CrossRef]

84. Jha, P.K.; Sadot, M.; Vinoa, S.A.; Jury, V.; Curet-Ploquina, S.; Rouaud, O.; Havet, M.; Le-Bail, A. A review on effect of DC voltage on crystallization process in food systems. *Innov. Food Sci. Emerg.* **2017**, *42*, 204–219. [CrossRef]

85. The potential energy of a continuous charge distribution is stored in the EF.

86. Han, B.; Chen, Z.; Louhi-Kultanen, M. Effect of a pulsed electric field on the synthesis of TiO_2 and its photocatalytic performance under visible light irradiation. *Powder Technol.* **2017**, *307*, 137–144. [CrossRef]

87. Garetz, B.A.; Myerson, A.S.; Arnold, S.; Aber, J.E. Method for Using a Static Electric Field to Induce Crystallization and to Control Crystal Form. U.S. Patent 7,879,115 B2, 13 April 2004.

crystals

MDPI

Article

Effect of an External Electric Field on the Kinetics of Dislocation-Free Growth of Tetragonal Hen Egg White Lysozyme Crystals

Haruhiko Koizumi*, Satoshi Uda, Kozo Fujiwara, Junpei Okada and Jun Nozawa

Institute for Materials Research, Tohoku University, 2-1-1 Katahira, Aoba-ku, Sendai 980-8577, Japan; uda@imr.tohoku.ac.jp (S.U.); kozo@imr.tohoku.ac.jp (K.F.); junpei.t.okada@imr.tohoku.ac.jp (J.O.); nozawa@imr.tohoku.ac.jp (J.N.)
* Correspondence: h_koizumi@imr.tohoku.ac.jp; Tel.: +81-22-215-2103; Fax: +81-22-215-2101

Received: 28 April 2017; Accepted: 9 June 2017; Published: 10 June 2017

Abstract: Dislocation-free tetragonal hen egg white (HEW) lysozyme crystals were grown from a seed crystal in a cell. The rates of tetragonal HEW lysozyme crystal growth normal to the (110) and (101) faces with and without a 1-MHz external electric field were measured. A decrease in the typical growth rates of the crystal measured under an applied field at 1 MHz was observed, although the overall driving force increased. Assuming that the birth and spread mechanism of two-dimensional nucleation occurs, an increase in the effective surface energy of the step ends was realized in the presence of the electric field, which led to an improvement in the crystal quality of the tetragonal HEW lysozyme crystals. This article also discusses the increase in the effective surface energy of the step ends with respect to the change in the entropy of the solid.

Keywords: protein crystals; growth kinetics; electric field; crystal quality

1. Introduction

The three-dimensional (3D) structures of protein molecules are closely related to the proteins found in living organisms. Thus, determination of the 3D structures of protein molecules is important for the advancement of medical science [1]. The 3D structures of protein molecules have typically been determined at synchrotron radiation facilities with high brilliance. A structure determined using data finer than 1.5 Å—which corresponds to the length of a covalent carbon-carbon bond—is needed in order to achieve structure-guided drug design and controlled drug delivery. This means that the collection of many high-order reflections of protein crystals (i.e., high diffraction efficiency — diffractivity) is desired to obtain accurate 3D structures of protein molecules. As such, many researchers have focused on developing a high-brilliance source and/or improving the sensitivity of the detector system [2]. However, useful structural analysis (<1.5 Å) of protein molecules for structure-guided drug design and controlled drug delivery represents only 9% of all protein molecules registered with the Protein Data Bank (PDB; http://www.rcsb.org/pdb/), even using synchrotron radiation facilities with high brilliance, such as SPring-8. This suggests the importance of the growth of high-quality protein crystals that allow the collection of many high-order reflections.

The growth of high-quality single crystals of proteins has been intensively pursued using magnetic fields [3–9], microgravity [10–16], solution flow [17–20], and gel as a growth host media [21–28]. A growth technique mediated by screw dislocations under low supersaturation has also been proposed [29–32]. Techniques in which an electric field is applied to a protein solution have also been actively investigated, with a focus on controlling the nucleation rate [33–49].

We have also previously attempted to control the nucleation process of protein crystals under an applied electrostatic field [50,51] by considering the effect of the field on the nucleation rate from a thermodynamics perspective. The effect of such a field was attributed to the electrostatic energy added to the chemical potentials of the liquid and solid phases. The electrostatic energy was produced by a large electric field of about 10^4 V/cm associated with an electric double layer (EDL) at the interface between the two phases, which was significantly larger than the actual experimentally-applied field (800 V/cm) [52]. This thermodynamic effect is added not only to the chemical potential, but also to the entropy. In particular, it was thermodynamically determined that the entropy of the solid decreases under an external electric field with a frequency of 1 MHz, and this is expected to result in a decrease in the degree of disorder in the crystal. Therefore, we attempted to improve the crystal quality of tetragonal hen egg white (HEW) lysozyme crystals by applying a 1-MHz external electric field. It was found that the full width at half-maximum (FWHM) of the X-ray diffraction rocking curves for the resulting crystals was lower than that in the absence of an electric field [53]. In addition, for crystals grown in an electric field, the FWHM was almost independent of the order of the diffraction peaks, whereas in the absence of an applied field, the FWHM increased for diffraction peaks with an order higher than 440 reflection [53], which suggests an improvement in local crystal quality. It was also found that the crystal homogeneity was improved under an electric field [54]. The improved crystal quality was attributed to a decrease in the misorientation between subgrains in the crystal [55]. The mechanism involved is not yet fully understood, although it may be related to the incorporation of impurities into the growing crystal.

In situ observation is a powerful tool for investigating the kinetics of protein crystal growth. The incorporation of impurities into the growing crystal has a significant effect on the kinetics of crystal growth. Therefore, the behavior of impurities can be understood by observing the change in crystal growth kinetics. If an external electric field has an effect on impurity incorporation, it would therefore also change the crystal growth kinetics. In this article, we reveal the effect of an external electric field on the incorporation of impurities into a growing crystal with respect to the change in crystal growth kinetics.

2. Experimental Procedure

HEW lysozyme was purchased from Wako Pure Chemical Industries, Ltd. (Osaka, Japan). There is typically a large amount of NaCl present in commercial HEW lysozyme; therefore, NaCl was removed from the HEW lysozyme solutions by dialysis. Tetragonal HEW lysozyme crystals grown from seed crystals were used in this work. Seed crystals were first grown by preparing an 80 mg/mL HEW lysozyme solution and a 1.0 M NaCl in 100 mM sodium acetate buffer solution and mixing them in equal volumes. The solutions were passed through a filter with a pore size of 0.20 μm to remove any foreign particulates or large protein aggregates. The resulting solution consisted of 40 mg/mL HEW lysozyme and 0.5 M NaCl in 100 mM sodium acetate buffer at pH 4.5. The seed crystals were grown from this crystallization solution at 21 °C for 1 day via the hanging drop technique. The grown seed crystals were chemically fixed by a modified version of a previously reported method [56,57]. The solution used for chemical cross-linking was a mixture of 2.5 wt % glutaraldehyde and 0.5 M NaCl in 100 mM sodium acetate buffer. The seed crystals were immersed in the cross-linking solution for 15 min at 23 °C. Therefore, no dislocations were introduced from the seed crystal, although it is possible that dislocations could occur during the growth process after two weeks [58]. This means that no dislocations occur during in situ observation, because the observation time is about one day. After cross-linking, the seed crystals were rinsed and reused because they did not dissolve in the undersaturated solution.

Next, a 100 mg/mL solution of HEW lysozyme and a solution of 1.0 M NaCl in 100 mM sodium acetate buffer were prepared and mixed in equal volumes. The resulting solution consisted of 50 mg/mL HEW lysozyme and 0.5 M NaCl in 100 mM sodium acetate buffer at pH 4.5. A tetragonal HEW lysozyme crystal was grown from a cross-linked seed crystal in a growth cell ($25 \times 25 \times 4$ mm). Crystal growth was conducted with and without a 1-MHz electrostatic field, and in situ observations were made using a digital microscope. Figure 1 shows a schematic illustration of the electrode arrangement on both sides of the growth cell. As shown in Figure 1, no current flows into HEW lysozyme solutions.

The electrodes were parallel to the ($\bar{1}$10) face of the seed crystal. Observations were made of the (110) face perpendicular to the electric field , in addition to the (101) face inclined by approximately 18° to the field.

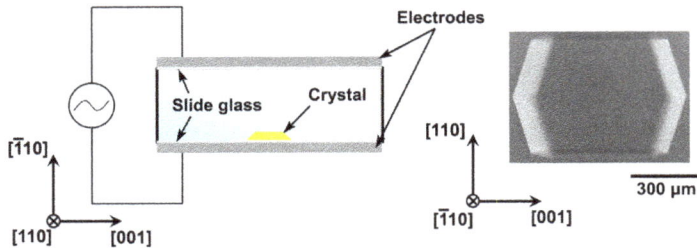

Figure 1. Schematic illustration of the experimental arrangement with electrodes on both sides of the growth cell.

The electric field strength was 1100 V/cm, and indium tin oxide-coated glass slides with a surface resistivity of 8–12 Ω/sq were used as the electrodes. However, in such a situation, it is expected that a large electric field of about 10^4 V/cm would be generated by the EDL at the interface between the solution and the crystal. The growth rate R for the crystals was measured in the presence and absence of the applied field, and the supersaturation σ ($=ln\frac{C}{C_{eq}}$—where C is the concentration of the solution and C_{eq} is the solubility) of the HEW lysozyme solution—was changed by varying the temperature. That is, all growth rates were measured using the same seed crystal at various temperatures. The supersaturation was estimated from the data given in Reference [59].

3. Results

Figure 2 shows the crystal dimensions along the [110] and [001] directions as a function of growth time with and without an external electric field, for a growth temperature at 18 °C.

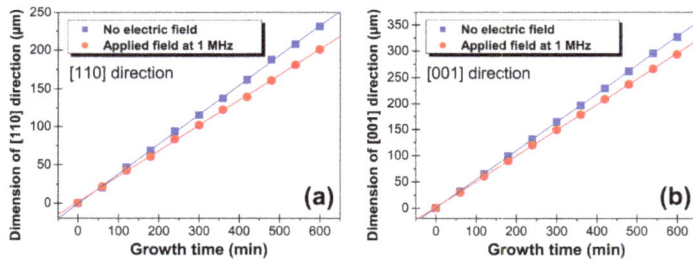

Figure 2. Crystal dimensions along (**a**) [110] and (**b**) [001] directions as function of growth time with and without external electric field, for growth temperature of 18 °C.

It can be seen that for both directions, the dimensions increase linearly with growth time, which indicates that the driving force is unchanged during the measurements. Additionally, the growth rate R, obtained from the slope of the straight lines, is smaller in the case of the applied field. R for the (110) and (101) faces can be calculated from the slope as follows:

$$R_{(110)} = \frac{g_{[110]}}{2},$$

(1)

$$R_{(101)} = \frac{g_{[001]}}{2} cos\theta, \tag{2}$$

where $g_{[110]}$ and $g_{[001]}$ are the slopes obtained for the [110] and [001] directions, respectively, and θ (=25.6°) is the angle between the normal to the (101) surface and the [001] direction. According to Eqations (1) and (2), R for the (110) and (101) faces in an applied field are 9.98 ± 0.06 μm/h and 13.26 ± 0.05 μm/h, respectively, while those in the absence of a field are 11.63 ± 0.07 μm/h and 14.76 ± 0.03 μm/h, respectively. Since the errors involved in these measurements were very small, it can be clearly concluded that the application of an electric field caused a reduction in the growth rate.

Figure 3 shows the dependence of the growth rate for the (110) and (101) faces on the degree of supersaturation. It can be seen that regardless of the supersaturation, the growth rate was always lower in the presence of an electric field. Again, the error bars for these measurements are very small. We have previously demonstrated that the nucleation rate during growth of tetragonal HEW lysozyme crystals increases (i.e., the driving force for nucleation increases) under an external electric field with a frequency of 1 MHz [50]. It might therefore be expected that the growth rate would also increase, in contrast to the results obtained in the present study. This would imply that the growth kinetics are changed under the applied field. We have previously observed that an external electric field does not affect the growth kinetics of YBCO superconductive oxides [60,61]. Therefore, this effect may be unique to protein crystals.

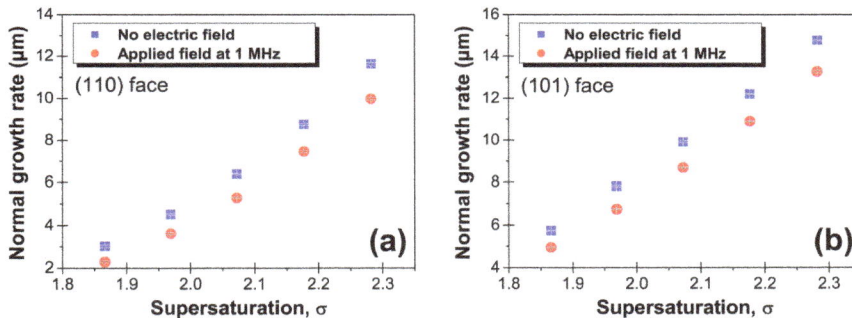

Figure 3. Supersaturation dependence of growth rate for (**a**) (110) and (**b**) (101) faces of tetragonal hen egg white (HEW) lysozyme crystals grown with and without an external electric field.

A dislocation-free crystal cross-linked for 15 min [58] was used in this experiment, so that 2D nucleation and growth of the birth-and-spread type should only occur on the growing faces. Under the range of supersaturations used in this experiment, the formation of multiple nuclei at several points on one growing face (multiple nucleation mode) has been observed for tetragonal HEW lysozyme crystals [62]. The growth rate R using the birth-and-spread model is expressed as [63,64]:

$$R = h(v_{step}^2 J)^{1/3} = h[(\Omega\beta_{step}(C - C_{eq}))^2 J]^{1/3}, \tag{3}$$

where J is the rate of two-dimensional (2D) nucleus formation, v_{step} is the tangential step velocity (which is related to the step kinetic coefficient β_{step} as $v_{step} = \Omega\beta_{step}(C - C_{eq})$ [65]), Ω is the kink volume, and h is the step height.

In the case of the (110) faces of tetragonal HEW lysozyme crystals, an elementary step involves two molecules [66–69]. The rate of 2D nucleus formation J on the (110) face can thus be expressed using a model derived from classical nucleation theory, as follows [62,70]:

$$J = \omega \Gamma Z exp(-\frac{\pi s \kappa^2}{2 k_B^2 T^2 \sigma}), \tag{4}$$

where ω is the frequency of attachment of molecules to the critical 2D nucleus, Γ is the Zeldovich factor, Z is the steady-state admolecule surface concentration, s is the area in which one molecule occupies inside the critical 2D nucleus ($s = 1.06 \times 10^{-17}$ m^2) [69], κ is the specific edge free energy, k_B is the Boltzmann constant, and T is the absolute temperature.

On the other hand, in the case of the (101) faces, an elementary step involves one molecule [66–69]; therefore, the rate of 2D nucleus formation J on the (101) face can be expressed according to the classical nucleation theory [62,70]:

$$J = \omega \Gamma Z exp(-\frac{\pi s \kappa^2}{k_B^2 T^2 \sigma}). \tag{5}$$

Here $s = 7.84 \times 10^{-18}$ m^2 [69].

Substituting Equation (4) into Equation (3) gives the growth rate in the coordinates $ln(R/\sigma^{1/6}(1 - e^{-\sigma})^{2/3})$ versus $1/\sigma T^2$ on the (110) face as follows:

$$ln(\frac{R_{(110)}}{\sigma^{1/6}(1 - e^{-\sigma})^{2/3}}) = A - \frac{\pi \Omega \alpha^2 h}{6 k_B^2} \frac{1}{\sigma T^2}, \tag{6}$$

where α is the effective surface free energy of the step end ($\alpha = \kappa/h$). Here, the step height on the (110) face is taken to be 5.6 nm [62].

In contrast, substituting Equation (5) into Equation (3) gives the growth rate in the coordinates $ln(R/\sigma^{1/6}(1 - e^{-\sigma})^{2/3})$ versus $1/\sigma T^2$ on the (101) face as follows:

$$ln(\frac{R_{(101)}}{\sigma^{1/6}(1 - e^{-\sigma})^{2/3}}) = A - \frac{\pi \Omega \alpha^2 h}{3 k_B^2} \frac{1}{\sigma T^2}. \tag{7}$$

Here, the step height of the (101) face is taken to be 3.4 nm [62].

Figure 4 shows the dependence of the growth rate R with and without an external electric field on $1/(\sigma T^2)$. The effective surface free energy for the step end, α, for the (110) and (101) faces can be estimated using Equations (6) and (7), respectively. Based on the straight line fit to the data in Figure 4a, α with and without an external electric field was estimated to be 1.249 mJ/m^2 and 1.189 mJ/m^2, respectively, for the (110) face. Table 1 shows the effective surface energy for the step end for the (110) and (101) faces with and without an external electric field.

Table 1. Effective surface energy for the step end α for the (110) and (101) faces of tetragonal HEW lysozyme crystals grown with and without an external electric field.

	Surface Energy, α (mJ/m^2)	
	(110) Face	**(101) Face**
No electric field	1.189 ± 0.009	1.329 ± 0.037
Applied field at 1 MHz	1.249 ± 0.023	1.364 ± 0.027

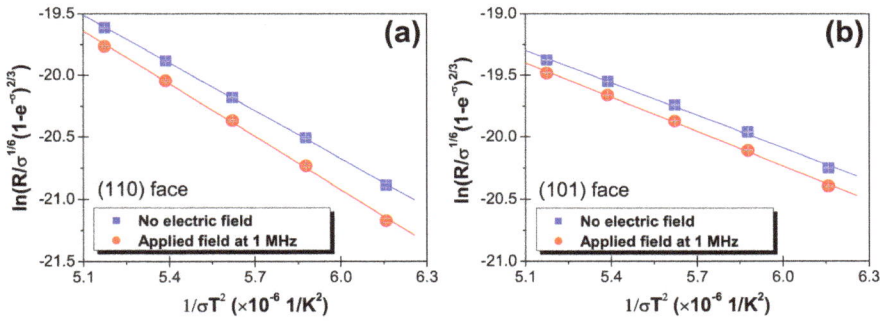

Figure 4. Growth rate R with and without external electric field as function of $1/(\sigma T^2)$. (**a**) (110) face and (**b**) (101) face.

For both faces, α is seen to be approximately 5% larger in the presence of the electric field. This would makes steps more difficult to form, thus leading to a flatter crystal surface. This could also prevent impurities from being incorporated into the steps during crystal growth, which would result in a decrease in the misorientation between subgrains. Therefore, the improved crystal quality for tetragonal HEW lysozyme crystals under a 1 MHz applied field may be predominantly caused by the increase in the effective surface energy of the step ends.

4. Discussion

Let us thermodynamically consider the increase in the effective surface energy of the step ends under an applied field. During nucleation, electrostatic energy is added to the chemical potentials of the liquid and solid phases by the electric field, which leads to an increase in the driving force for nucleation [50]. This would seem to suggest that the growth velocity would also increase. However, in the present study, the electric field was found to cause a reduction in the growth rate. Thus, the increase in the effective surface energy of step ends due to the electrostatic energy may overcome the increase in the driving force for nucleation. Employing the Helmholtz free energy, the free energy on the crystal surface F_s can be expressed as:

$$F_s = U_s - TS_s, \tag{8}$$

where U_s is the energy required for formation of a step and S_s is the entropy related to the shape of the step. Therefore, the electrostatic energy added to the energy required for step formation and the entropy related to the shape of the step must be considered. The modified energy due to formation of the step $U_{s(E)}$, and the modified entropy related to a shape of the step $S_{s(E)}$, can be derived from the Helmholtz free energy:

$$U_{s(E)} = U_{s(0)} + \frac{1}{2}V_c E^2 [\epsilon - T\frac{\partial \epsilon}{\partial T}], \tag{9}$$

$$S_{s(E)} = S_{s(0)} - \frac{1}{2}V_c E^2 \frac{\partial \epsilon}{\partial T}, \tag{10}$$

where $U_{s(0)}$ and $S_{s(0)}$ are the energy required for formation of the step and the entropy related to the shape of the step without an external electric field, respectively, T is the absolute temperature, ϵ is the electrical permittivity, E is the strength of the external electric field, and V_c is the volume to which the external electric field is applied. Therefore, the free energy on the crystal surface would increase if the energy required to form the step increases or the entropy related to the shape of the step decreases under an external electric field.

First, let us consider the effect of the external electric field on the entropy. The dependence of the electrical permittivity of protein crystals on the temperature has been measured using monoclinic lysozyme crystals [71], whereby the sign of the derivative for protein crystals is positive (i.e., the entropy related to the shape of the step decreases under an external electric field). On the other hand, whether the energy required to form the step increases or decreases is attributed to the magnitude between the electrical permittivity and the temperature dependence of the electrical permittivity, as seen in Equation (9). The temperature dependence of the electrical permittivity for protein crystals can be estimated to be about 10^{-11} C^2/Nm^2K using the data reported by Rashkovich et al. [71], whereas the electrical permittivity for protein crystals has been measured to be about 10^{-10} C^2/Nm^2 [71]. Thus, $\epsilon - T\frac{\partial \epsilon}{\partial T}$ is negative, which leads to a reduction of the energy required to form a step under an external electric field. Thus, under an applied field, the increase in the effective surface energy of the step end could be caused by a decrease in the entropy related to the shape of the step. That is, the improvement in crystal quality under an electric field could be due to a decrease in the entropy related to the shape of the step.

In addition, this result could indicate that control of the effective surface energy of the step ends would play an important role in the growth of high-quality protein crystals, which could be achieved by a decrease in the entropy related to the shape of steps and/or an increase in the energy required to form steps.

Acknowledgments: This work was supported in part by a Grant-in-Aid for Challenging Exploratory Research (No. 15K14484) from the Ministry of Education, Culture, Sports, Science and Technology of Japan.

Author Contributions: The study was designed by Haruhiko Koizumi; The experimental results were discussed by all authors.

Conflicts of Interest: The authors declare no conflicts of interests.

References

1. Kuhn, P.; Wilson, K.; Patch, M.G.; Stevens, R.C. The genesis of high-throughput structure-based drug discovery using protein crystallography. *Curr. Opin. Chem. Biol.* **2002**, *6*, 704–710.
2. Chayen, N.E.; Helliwell, J.R.; Snell, E.H. *Macromolecular Crystallization and Crystal Perfection*; Oxford University Press: Oxford, UK, 2010.
3. Sato, T.; Yamada, Y.; Saijo, S.; Hori, T.; Hirose, R.; Tanaka, N.; Sazaki, G.; Nakajima, K.; Igarashi, N.; Tanaka, M.; et al. Enhancement in the perfection of orthorhombic lysozyme crystals grown in a high magnetic field (10 T). *Acta Crystallogr.* **2000**, *D56*, 1079–1083.
4. Lin, S.; Zhou, M.; Azzi, A.; Xu, G.; Wakayama, N.; Ataka, M. Magnet used for protein crystallization: Novel attempts to improve the crystal quality. *Biochem. Biophys. Res. Commun.* **2000**, *275*, 274–278.
5. Ataka, M.; Wakayama, N. Effects of a magnetic field and magnetization force on protein crystal growth. Why does a magnet improve the quality of some crystals? *Acta Crystallogr.* **2002**, *D58*, 1708–1710.
6. Wakayama, N. Effects of a strong magnetic field on protein crystal growth. *Cryst. Growth Des.* **2003**, *3*, 17–24.
7. Kinoshita, T.; Ataka, M.; Warizaya, M.; Neya, M.; Fujii, T. Improving quality and harvest period of protein crystals for structure-based drug design: effects of a gel and a magnetic field on bovine adenosine deaminase crystals. *Acta Crystallogr.* **2003**, *D59*, 1333–1335.
8. Lübbert, D.; Meents, A.; Weckert, E. Accurate rocking-curve measurements on protein crystals grown in a homogeneous magnetic field of 2.4 T. *Acta Crystallogr.* **2004**, *D60*, 987–998.
9. Moreno, A.; Quiroz-García, B.; Yokaichiya, F.; Stojanoff, V.; Rudolph, P. Protein crystal growth in gels and stationary magnetic fields. *Cryst. Res. Technol.* **2007**, *42*, 231–236.
10. DeLucas, L.; Smith, C.; Smith, H.; Vijay-Kumar, S.; Senadhi, S.; Ealick, S.; Carter, D.; Snyder, R.; Weber, P.; Salemme, F. Protein crystal growth in microgravity. *Science* **1989**, *246*, 651–654.
11. McPherson, A. Virus and protein crystal growth on earth and in microgravity. *J. Phys. D* **1993**, *26*, 104–112.

12. Snell, E.; Weisgerber, S.; Helliwell, J.; Weckert, E.; Holzer, K.; Schroer, K. Improvements in lysozyme protein crystal perfection through microgravity growth. *Acta Crystallogr.* **1995**, *D51*, 1099–1102.

13. Sato, M.; Tanaka, H.; Inaka, K.; Shinozaki, S.; Yamanaka, A.; Takahashi, S.; Yamanaka, M.; Hirota, E.; Sugiyama, S.; Kato, M.; et al. JAXA-GCF project-high-quality protein crystals grown under microgravity environment for better understanding of protein structure. *Microgravity Sci. Technol.* **2006**, *18*, 184–189.

14. Takahashi, S.; Tsurumura, T.; Aritake, K.; Furubayashi, N.; Sato, M.; Yamanaka, M.; Hirota, E.; Sano, S.; Kobayashi, T.; Tanaka, T.; et al. High-quality crystals of human haematopoietic prostaglandin D synthase with novel inhibitors. *Acta Crystallogr.* **2010**, *F66*, 846–850.

15. Inaka, K.; Takahashi, S.; Aritake, K.; Tsurumura, T.; Furubayashi, N.; Yan, B.; Hirota, E.; Sano, S.; Sato, M.; Kobayashi, T.; et al. High-quality protein crystal growth of mouse lipocalin-type prostaglandin D synthase in microgravity. *Cryst. Growth Des.* **2011**, *11*, 2107–2111.

16. Yoshikawa, S.; Kukimoto-Niino, M.; Parker, L.; Handa, N.; Terada, T.; Fujimoto, T.; Terazawa, Y.; Wakiyama, M.; Sato, M.; Sano, S.; et al. Structural basis for the altered drug sensitivities of non-small cell lung cancer-associated mutants of human epidermal growth factor receptor. *Oncogene* **2012**, *32*, 27–38.

17. Vekilov, P.G.; Thomas, B.R.; Rosenberger, F. Effects of convective solute and impurity transport in protein crystal growth. *J. Phys. Chem. B* **1998**, *102*, 5208–5216.

18. Kadowaki, A.; Yoshizaki, I.; Adachi, S.; Komatsu, H.; Odawara, O.; Yoda, S. Effects of forced solution flow on protein-crystal quality and growth process. *Cryst. Growth Des.* **2006**, *6*, 2398–2403.

19. Otálora, F.; Gavira, J.A.; Ng, J.D.; García-Ruiz, J.M. Counterdiffusion methods applied to protein crystallization. *Prog. Biophys. Mol. Biol.* **2009**, *101*, 26–37.

20. Maruyama, M.; Kawahara, H.; Sazaki, G.; Maki, S.; Takahashi, Y.; Yoshikawa, H.Y.; Sugiyama, S.; Adachi, H.; Takano, K.; Matsumura, H.; et al. Effects of a forced solution flow on the step advancement on {110} faces of tetragonal lysozyme crystals: direct visualization of individual steps under a forced solution flow. *Cryst. Growth Des.* **2012**, *12*, 2856–2863.

21. Garcia-Ruiz, J.; Moreno, A. Investigations on protein crystal growth by the gel acupuncture method. *Acta Crystallogr.* **1994**, *D50*, 484–490.

22. Vidal, O.; Robert, M.; Arnoux, B.; Capelle, B. Crystalline quality of lysozyme crystals grown in agarose and silica gels studied by X-ray diffraction techniques. *J. Cryst. Growth* **1999**, *196*, 559–571.

23. Lorber, B.; Sauter, C.; Ng, J.; Zhu, D.; Giegé, R.; Vidal, O.; Robert, M.; Capelle, B. Characterization of protein and virus crystals by quasi-planar wave X-ray topography: A comparison between crystals grown in solution and in agarose gel. *J. Cryst. Growth* **1999**, *204*, 357–368.

24. Dong, J.; Boggon, T.J.; Chayen, N.E.; Raftery, J.; Bi, R.C.; Helliwell, J.R. Bound-solvent structures for microgravity-, ground control-, gel-and microbatch-grown hen egg-white lysozyme crystals at 1.8 A resolution. *Acta Crystallogr.* **1999**, *D55*, 745–752.

25. Garcıa-Ruiz, J.; Novella, M.; Moreno, R.; Gavira, J. Agarose as crystallization media for proteins: I: Transport processes. *J. Cryst. Growth* **2001**, *232*, 165–172.

26. Gavira, J.A.; García-Ruiz, J.M. Agarose as crystallisation media for proteins II: Trapping of gel fibres into the crystals. *Acta Crystallogr.* **2002**, *D58*, 1653–1656.

27. Sugiyama, S.; Maruyama, M.; Sazaki, G.; Hirose, M.; Adachi, H.; Takano, K.; Murakami, S.; Inoue, T.; Mori, Y.; Matsumura, H. Growth of protein crystals in hydrogels prevents osmotic shock. *JACS* **2012**, *134*, 5786–5789.

28. Maruyama, M.; Hayashi, Y.; Yoshikawa, H.Y.; Okada, S.; Koizumi, H.; Tachibana, M.; Sugiyama, S.; Adachi, H.; Matsumura, H.; Inoue, T.; et al. A crystallization technique for obtaining large protein crystals with increased mechanical stability using agarose gel combined with a stirring technique. *J. Cryst. Growth* **2016**, *452*, 172–178.

29. Sleutel, M.; Sazaki, G.; Van Driessche, A.E. Spiral-mediated growth can lead to crystals of higher purity. *Cryst. Growth Des.* **2012**, *12*, 2367–2374.

30. Sleutel, M.; Van Driessche, A.E. On the self-purification cascade during crystal growth from solution. *Cryst. Growth Des.* **2013**, *13*, 688–695.

31. Hayashi, Y.; Maruyama, M.; Yoshimura, M.; Okada, S.; Yoshikawa, H.Y.; Sugiyama, S.; Adachi, H.; Matsumura, H.; Inoue, T.; Takano, K.; et al. Spiral growth can enhance both the normal growth rate and quality of tetragonal lysozyme crystals grown under a forced solution flow. *Cryst. Growth Des.* **2015**, *15*, 2137–2143.

32. Tominaga, Y.; Maruyama, M.; Yoshimura, M.; Koizumi, H.; Tachibana, M.; Sugiyama, S.; Adachi, H.; Tsukamoto, K.; Matsumura, H.; Takano, K.; et al. Promotion of protein crystal growth by actively switching crystal growth mode via femtosecond laser ablation. *Nat. Photonics* **2016**, *10*, 723–726.

33. Taleb, M.; Didierjean, C.; Jelsch, C.; Mangeot, J.; Capelle, B.; Aubry, A. Crystallization of proteins under an external electric field. *J. Cryst. Growth* **1999**, *200*, 575–582.

34. Taleb, M.; Didierjean, C.; Jelsch, C.; Mangeot, J.; Aubry, A. Equilibrium kinetics of lysozyme crystallization under an external electric field. *J. Cryst. Growth* **2001**, *232*, 250–255.

35. Nanev, C.; Penkova, A. Nucleation of lysozyme crystals under external electric and ultrasonic fields. *J. Cryst. Growth* **2001**, *232*, 285–293.

36. Charron, C.; Didierjean, C.; Mangeot, J.; Aubry, A. TheOctopus' plate for protein crystallization under an electric field. *J. Appl. Crystallogr.* **2003**, *36*, 1482–1483.

37. Mirkin, N.; Frontana-Uribe, B.; Rodríguez-Romero, A.; Hernández-Santoyo, A.; Moreno, A. The influence of an internal electric field upon protein crystallization using the gel-acupuncture method. *Acta Crystallogr.* **2003**, *D59*, 1533–1538.

38. Moreno, A.; Sazaki, G. The use of a new ad hoc growth cell with parallel electrodes for the nucleation control of lysozyme. *J. Cryst. Growth* **2004**, *264*, 438–444.

39. Penkova, A.; Gliko, O.; Dimitrov, I.; Hodjaoglu, F.; Nanev, C.; Vekilov, P. Enhancement and suppression of protein crystal nucleation due to electrically driven convection. *J. Cryst. Growth* **2005**, *275*, 1527–1532.

40. Penkova, A.; Pan, W.; Hodjaoglu, F.; Vekilov, P. Nucleation of protein crystals under the influence of solution shear flow. *Ann. N. Y. Acad. Sci.* **2006**, *1077*, 214–231.

41. Al-Haq, M.; Lebrasseur, E.; Choi, W.; Tsuchiya, H.; Torii, T.; Yamazaki, H.; Shinohara, E. An apparatus for electric-field-induced protein crystallization. *J. Appl. Crystallogr.* **2007**, *40*, 199–201.

42. Al-Haq, M.; Lebrasseur, E.; Tsuchiya, H.; Torii, T. Protein crystallization under an electric field. *Crystallogr. Rev.* **2007**, *13*, 29–64.

43. Hammadi, Z.; Astier, J.; Morin, R.; Veesler, S. Protein crystallization induced by a localized voltage. *Cryst. Growth Des.* **2007**, *7*, 1472–1475.

44. Pérez, Y.; Eid, D.; Acosta, F.; Marín-García, L.; Jakoncic, J.; Stojanoff, V.; Frontana-Uribe, B.; Moreno, A. Electrochemically assisted protein crystallization of commercial cytochrome c without previous purification. *Cryst. Growth Des.* **2008**, *8*, 2493–2496.

45. Mirkin, N.; Jaconcic, J.; Stojanoff, V.; Moreno, A. High resolution X-ray crystallographic structure of bovine heart cytochrome c and its application to the design of an electron transfer biosensor. *Proteins Struct. Funct. Bioinf.* **2008**, *70*, 83–92.

46. Hou, D.; Chang, H. ac field enhanced protein crystallization. *Appl. Phys. Lett.* **2008**, *92*, 223902.

47. Revalor, E.; Hammadi, Z.; Astier, J.; Grossier, R.; Garcia, E.; Hoff, C.; Furuta, K.; Okustu, T.; Morin, R.; Veesler, S. Usual and unusual crystallization from solution. *J. Cryst. Growth* **2010**, *312*, 939–946.

48. Wakamatsu, T. Transparent cell for protein crystallization under low applied voltage. *Jpn. J. Appl. Phys.* **2011**, *50*, 048003.

49. Wakamatsu, T.; Toyoshima, S.; Shimizu, H. Observation of electric-field induced aggregation in crystallizing protein solutions by forward light scattering. *Appl. Phys. Lett.* **2011**, *99*, 153701.

50. Koizumi, H.; Fujiwara, K.; Uda, S. Control of nucleation rate for tetragonal hen-egg white lysozyme crystals by application of an electric field with variable frequencies. *Cryst. Growth Des.* **2009**, *9*, 2420–2424.

51. Koizumi, H.; Tomita, Y.; Uda, S.; Fujiwara, K.; Nozawa, J. Nucleation rate enhancement of porcine insulin by application of an external AC electric field. *J. Cryst. Growth* **2012**, *352*, 155–157.

52. Koizumi, H.; Fujiwara, K.; Uda, S. Role of the electric double layer in controlling the nucleation rate for tetragonal hen egg white lysozyme crystals by application of an external electric field. *Cryst. Growth Des.* **2010**, *10*, 2591–2595.

53. Koizumi, H.; Uda, S.; Fujiwara, K.; Tachibana, M.; Kojima, K.; Nozawa, J. Improvement of crystal quality for tetragonal hen-egg white lysozyme crystals under application of an external AC electric field. *J. Appl. Crystallogr.* **2013**, *46*, 25–29.

54. Koizumi, H.; Uda, S.; Fujiwara, K.; Tachibana, M.; Kojima, K.; Nozawa, J. Enhancement of crystal homogeneity of protein crystals under application of an external alternating current electric field. *AIP Conf. Proc.* **2014**, *1618*, 265–268.

55. Koizumi, H.; Uda, S.; Fujiwara, K.; Tachibana, M.; Kojima, K.; Nozawa, J. Control of subgrain formation in protein crystals by the application of an external electric field. *Cryst. Growth Des.* **2014**, *14*, 5662–5667.
56. Iimura, Y.; Yoshizaki, I.; Rong, L.; Adachi, S.; Yoda, S.; Komatsu, H. Development of a reusable protein seed crystal processed by chemical cross-linking. *J. Cryst. Growth* **2005**, *275*, 554–560.
57. Koizumi, H.; Tachibana, M.; Yoshizaki, I.; Fukuyama, S.; Tsukamoto, K.; Suzuki, Y.; Uda, S.; Kojima, K. Dislocations in high-quality glucose isomerase crystals grown from seed crystals. *Cryst. Growth Des.* **2014**, *14*, 5111–5116.
58. Koizumi, H.; Uda, S.; Tachibana, M.; Tsukamoto, K.; Kojima, K.; Nozawa, J. Crystallization technique for strain-free protein crystals using cross-linked seed crystals. *Cryst. Growth Des.* **2016**, *16*, 6089–6094.
59. Cacioppo, E.; Pusey, M.L. The solubility of the tetragonal form of hen egg white lysozyme from pH 4.0 to 5.4. *J. Cryst. Growth* **1991**, *114*, 286–292.
60. Huang, X.; Uda, S.; Yao, X.; Koh, S. In situ observation of crystal growth process of YBCO superconductive oxide with an external electric field. *J. Cryst. Growth* **2006**, *294*, 420–426.
61. Huang, X.; Uda, S.; Koh, S. Effect of an external electric field on the crystal growth process of YBCO superconductive oxide. *J. Cryst. Growth* **2007**, *307*, 432–439.
62. Van Driessche, A.E.; Sazaki, G.; Otálora, F.; González-Rico, F.M.; Dold, P.; Tsukamoto, K.; Nakajima, K. Direct and noninvasive observation of two-dimensional nucleation behavior of protein crystals by advanced optical microscopy. *Cryst. Growth Des.* **2007**, *7*, 1980–1987.
63. Malkin, A.; Chernov, A.; Alexeev, I. Growth of dipyramidal face of dislocation-free ADP crystals; free energy of steps. *J. Cryst. Growth* **1989**, *97*, 765–769.
64. Sleutel, M.; Willaert, R.; Gillespie, C.; Evrard, C.; Wyns, L.; Maes, D. Kinetics and thermodynamics of glucose isomerase crystallization. *Cryst. Growth Des.* **2008**, *9*, 497–504.
65. Vekilov, P.G.; Alexander, J.I.D. Dynamics of layer growth in protein crystallization. *Chem. Rev.* **2000**, *100*, 2061–2090.
66. Durbin, S.; Feher, G. Studies of crystal growth mechanisms of proteins by electron microscopy. *J. Mol. Biol.* **1990**, *212*, 763–774.
67. Durbin, S.D.; Carlson, W.E. Lysozyme crystal growth studied by atomic force microscopy. *J. Cryst. Growth* **1992**, *122*, 71–79.
68. Konnert, J.H.; D'Antonio, P.; Ward, K. Observation of growth steps, spiral dislocations and molecular packing on the surface of lysozyme crystals with the atomic force microscope. *Acta Crystallogr.* **1994**, *D50*, 603–613.
69. Li, H.; Nadarajah, A.; Pusey, M.L. Determining the molecular-growth mechanisms of protein crystal faces by atomic force microscopy. *Acta Crystallogr.* **1999**, *D55*, 1036–1045.
70. Markov, I.V. *Crystal Growth for Beginners: Fundamentals of Nucleation, Crystal Growth and Epitaxy*; World Scientific: Singapore, 2003.
71. Rashkovich, L.; Smirnov, V.; Petrova, E. Some dielectric properties of monoclinic lysozyme crystals. *Phys. Solid State* **2008**, *50*, 631–637.

crystals

MDPI

Article

Crystallization under an External Electric Field: A Case Study of Glucose Isomerase

Evgeniya Rubin [1,2], Christopher Owen [1,3] and Vivian Stojanoff [1,4,*

[1]   National Syncrhotron Light Source, Upton, NY 11973, USA; evgeniyarubin@gmail.com (E.R.);
      chowen15@gmail.com (C.O.)
[2]   Accenture, New York, NY 10105, USA
[3]   International Imaging Materials—IIMAK, Amherst, NY 14228, USA
[4]   National Syncrhotron Light Source II, Upton, NY 11973, USA
*   Correspondence: stojanof@bnl.gov; Tel.: +1-631-344-8375

Academic Editor: Abel Moreno
Received: 9 May 2017; Accepted: 9 June 2017; Published: 6 July 2017

Abstract: Electric fields have been employed to promote macromolecular crystallization for several decades. Although crystals grown in electric fields seem to present higher diffraction quality, these methods are not widespread. For most configurations, electrodes are in direct contact with the protein solution. Here, we propose a configuration that can be easily extended to standard crystallization methods for which the electrodes are not in direct contact with the protein solution. Furthermore, the proposed electrode configuration supplies an external DC electric field. Glucose Isomerase from *Streptomyces rubiginosus* crystals were grown at room temperature using the microbatch method in the presence of 1, 2, 4, and 6 kV. Several crystallization trials were carried out for reproducibility and statistical analysis purposes. The comparison with crystals grown in the absence of electric fields showed that crystallization in the presence of electric fields increases the size of crystals, while decreasing the number of nucleations. X-ray diffraction analysis of the crystals showed that those grown in the presence of electric fields are of higher crystal quality.

Keywords: crystallization; macromolecular crystallography; external DC electric field; microbatch method

1. Introduction

The growth of high quality macromolecular crystals has been addressed theoretically and experimentally by multiple studies. The nucleation and growth process of macromolecular crystals [1,2] depends on a variety of physical and chemical parameters. Essentially any crystallization method aims to drive a solution from an undersaturated condition to a supersaturation state. As macromolecules nuclei are formed and crystals start growing, the concentration of macromolecules in the solution decreases, eventually becoming equal to the solubility, and consequently leading to the cessation of the crystal growth process. Crystal growth rates typically respond to concentration changes at the growing interface and therefore respond to a combination of transport phenomena, such as the diffusion of chemical species and interface kinetics [3]. Several crystallization methods aim to control the former, since any inhomogeneities in the growth solution will lead to convective flow and sedimentation. These may be minimized under low gravity conditions [4,5]. The relatively small diffusion coefficient observed for macromolecules [6] may enhance the probability for uniform incorporation of the molecules at the growing interface. Similarly, the use of gels [7,8], lipidic cubic phase [9,10], or oils in near containerless systems [11] reduce convection and sedimentation by slowing down diffusive transport of the macromolecules. Growth kinetics has been mostly controlled through supersaturation and nucleation inductors, for example through seeding [12–14].

Electromagnetic fields have been employed to control growth rates, and specific magnetic fields have been employed to minimize buoyancy-driven convection and sedimentation [15,16] and have in turn improved crystal quality [17]. Internal and external electric fields have been applied to improve the macroscopic and microscopic quality of macromolecular crystals; for a review see, for example, Hammadi et al. [18,19]. Probably the first macromolecule to crystallize in an external DC electric field was Estradiol 17b-dehydrogenase from human placenta, in a study conducted by Chin et al. [20]. They obtained crystals by applying an increasing voltage from 100 to 300 kV over 24–48 h and estimated 1500 V/m electric field using a method they had devised, electrophoresis diffusion. However, hen egg white lysozyme (HEWL) is the macromolecule of choice to study the effect of electric fields on the nucleation and growth process. Adapting two flat electrodes to the usual hanging and sitting drop setup, Taleb and co-workers [21,22] reported that the largest effect was observed for the condition with the highest positive charge, 11 (pH = 4.5 and high NaCl concentrations (1 M) [23], and an external electrical field intensity, E = 7.5×10^4 V/m. Larger field intensities, 1.5×10^5 V/m, were reported by Nanev and Penkova [24,25], who developed a quasi 2D glass cell to crystallize HEWL and simultaneously control temperature and reduce convection. Garetz et al. [26] reported the effect of laser light on the improved crystallization quality of glycine crystals. They attributed the observed effect to the small dipole moment induced by an electric field of a laser light flash. They argued that the electric field induced by the laser light lowers the molecule energy, favoring molecular orientation along the polarization of the laser electric field. They attributed the effect to the molecule's polarizability, which is acted on by the electric field of the laser light. A small dipole moment was induced and the molecule's energy was lowered, favoring molecular orientation. Appropriate field intensities would induce electrically-driven solution flow [27], promoting molecules to align and consequently leading to crystallization. In spite of the improved crystal quality reported by the crystallization of proteins under the influence of electric fields, the use of electric fields to control the kinetics of protein crystallization is not widespread. One possible reason for this might be the necessity of special settings and the usage of large quantities of protein. Two types of settings are commonly used. In the first, the electrodes are in contact with the growth solution [28–33], while in the second, electrodes are placed externally [24,25,34–37].

Here, we explore the effect of external electric fields on the crystallization process of Glucose Isomerase (GI) from *Streptomyces rubiginosus*, an enzyme used industrially to convert glucose to fructose in the manufacturing of high-fructose corn syrup. We take advantage of the microbatch method [38,39] that allows for low protein consumption, and combine it with a unique design of electrodes that allow for flexibility and the simultaneous application of electric fields of different intensities. About 360 trials for each of the five electric field conditions studied were analyzed, and approximately 75 crystals were exposed to X-rays for quality assessment. Our results indicate that the quality of the crystals obtained increased, while the number of nucleation sites decreased with the increase of the electric field applied.

2. Results

A total five experiments were carried out. For each experiment, five microbatch plates were prepared and respectively submitted to 0, 1, 2, 4, and 6 kV. Since each plate has 72 wells, 360 identical conditions were analyzed. For each trial, the number of crystals were recorded for the corresponding voltage, as shown in Figure 1. As the applied voltage increased, the number of nucleation centers decreased. A Poisson analysis of the number of nucleations per well as a function of electric field intensity or applied voltage shows that the least number of nucleations, i.e., the number of crystals per well, was observed for 6 kV, with less than one crystal per trial ($\lambda = 0.75$; where λ is the average number of crystals per trial). If crystallization takes place in the absence of an applied voltage, 6 to 7 crystals were expected per trial ($\lambda = 6.62$). The typical crystal size distribution is shown in Figure 2. The mean size distribution indicates that with an increase in the applied voltage, the size of the crystals also increased. The mean value chart shown in Figure 2b indicates that plates submitted to 4 and 6 kV were most likely to have crystals of 400 to 500 μm in size.

All Glucose Isomerase crystals analyzed belonged to the I222 space group, independent of the electric field applied. Crystal quality parameters were assessed from the X-ray diffraction patterns for three of the five experiments, due to beam time constraints. Parameters such as mosaicity, resolution, and signal-to-noise ratio ($I/\sigma I$) were recorded for five crystals of similar dimensions randomly chosen from each crystallization plate and analyzed at room temperature. Mosaicity is a measure of disorder in a crystal. The lower the mosaicity, the more organized the molecular packing in the crystal. Lower mosaicity values indicate increased crystalline organization [17–40], that is reflected in an improved crystal structure determination. The lower mosaicity values observed for crystals grown under the influence of an electric field in comparison to the mosaicity determined for crystals grown in the absence of an electric field is an indication of the influence of electric fields on the nucleation and crystallization process. As the applied voltage increased, the mosaicity was observed to decrease, as shown in Figure 3a. Diffraction resolution is related to the smallest measurable distance between atoms in a crystal. The resolution values of the crystals grown in an electric field were lower than those determined for crystals grown without an electric field, as illustrated in Figure 3b. This finding indicates that the increase in voltage improved the diffraction resolution, giving the possibility of a more detailed structure determination and a better understanding of the molecular function.

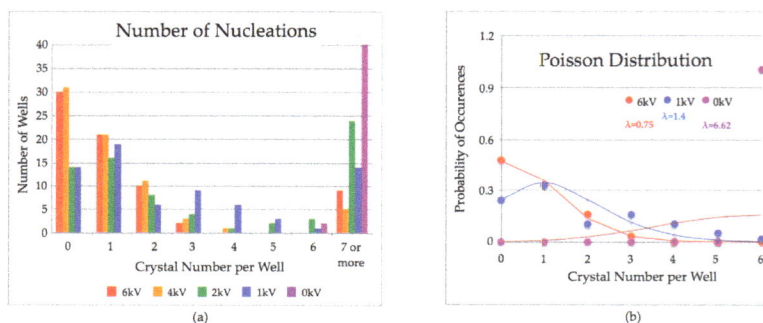

Figure 1. Distribution of the number of Glucose Isomerase crystals grown by the microbatch method in the presence of an external DC electric field for 48 h. In total, 360 microbatch wells were probed. (**a**) Number of nucleation sites for the five voltages probed; (**b**) Poisson distribution of the number of crystals per well indicates that the optimal electric field for crystallization of Glucose Isomerase lies around 5 kV.

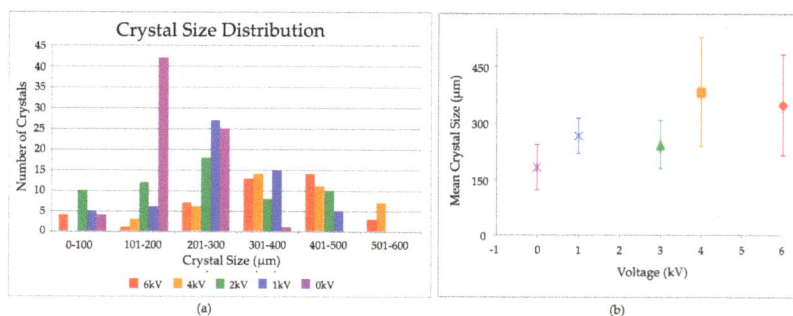

Figure 2. Size distribution of Glucose Isomerase crystals grown by the microbatch method in the presence of an external DC electric field for 48 h. In total, 360 micro batch wells were probed. (**a**) Crystal size for the five voltages probed; (**b**) Mean crystal size as a function of applied voltage. The largest crystals were obtained for DC voltages between 4 and 6 kV.

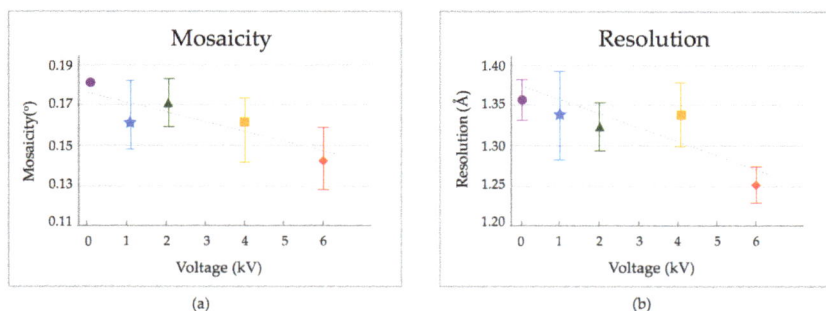

Figure 3. X-ray crystal quality analysis of Glucose Isomerase crystals grown by the microbatch method in the presence of an external DC electric field for 48 h. In total, 75 crystals grown by the microbatch method were probed; five crystals for each field condition. (**a**) Average mosaicity; (**b**) Average diffraction resolution. Color scheme follows that of previous figures; red 6 keV, orange 4 keV, green 2 keV, blue 1 keV and purple 0 keV.

3. Discussion

Improved crystal quality for macromolecules crystallized in the presence of internal and external electric fields has been reported by several authors. Initial crystallization trials conducted with the Efield microbatch device were performed using HEWL (results not published). The trials included the four DC fields, 1, 2, 4, 6 kV, as well as crystallization plates not submitted to an external electric field that served as a control. HEWL crystals grew larger in size and the number of crystals per well diminished for increasing electric field intensities. Similar results were reported by other authors [18–22,24,25,30–37], including reports that HEWL crystals formed preferentially close to the oil-growth solution interface [34]. Crystal alignment with the electric field was also observed [28–37]. X-ray diffraction analysis confirmed the increase in crystal quality with increasing field intensity [19]. Rocking curve measurements on different crystallographic planes, reported by the Koizumi's group [35,36], showed the differential effect of the electric field. These measurements require specific experimental arrangements and are not available on the standard macromolecular crystallography synchrotron beam lines, but can provide valuable information on the growth process.

Glucose Isomerase from *Streptomyces rubiginosus* was crystalized in the Efield microbatch device in the presence of four different electric field intensities (applied DC voltages: 1, 2, 4, 6 kV) and, for a control, in the absence of a field. The intensity of the electric field provided by the setup according to the applied voltage was estimated with the help of Autodesk Inventor 2013 (https://www.autodesk.com/education/free-software/inventor-professional) and Ansoft Maxwell 3D version 12 software (http://ansoft-maxwell.narod.ru/en/CompleteMaxwell3D_V11.pdf). The results of these simulations were viewed in cross-section and included the finite element mesh used by the Maxwell 3D software: the voltage, the magnitude of the electric field, and the electric field vector lines within the model. The estimated electric field varied from ~10^2 V/m to ~10^3 V/m for the DC voltages applied. An example of the field intensity at different positions in the microbatch plate is shown in Figure 4 for a single well with an applied DC voltage of 6 kV.

Light microscopy analysis of GI crystals showed a decrease in the nucleation rate and an increase in crystal size with the increase in applied voltage. Poisson distribution analysis of the number of crystals per trial, shown in Figure 1b, indicates that the best response to an electric field would be achieved with applied voltages around 5 kV. The microbatch Efield device has been designed to supply pre-fixed voltages of 1, 2, 4, and 6 kV. To be able to select different DC voltages for the experiment would require a design modification, which is being pursued. X-ray diffraction analysis indicated that crystals grown in the presence of an electric field have lower mosaicity and present higher diffraction resolution [19–36], showing that the crystallization process improved in the presence of electric fields. The field intensity of the order of 3×10^3 V/m is in accord with the field intensity estimated by Chin et al. [20], 1.5×10^3 V/m, for the crystallization of human placental estradiol 17beta-dehydrogenase. However, it is three or four orders of magnitude smaller than those reported for HEWL crystallization in the presence of an external electric field ~10^6–10^7 V/m [18,24,34].

The electric dipole moment of Glucose Isomerase from *Streptomyces rubiginosus* was determined using the crystal structure reported in the protein data bank (PDB ID 4A8L) as input to the online software provided by the Weizman Institute, Israel (http://dipole.weizmann.ac.il) [41]. The monomer showing the electric dipole moment vector and surface charge (Coot: http://lmb.bioch.ox.ac.uk/coot/) [42] are represented in Figure 4. The electric dipole moment for other macromolecules frequently crystallized in the presence of an external electric field was also determined (Table 1). The dipole moment of HEWL (PDB ID 193L), 198D, is about 10 times smaller than the dipole moment determined for Glucose Isomerase (PDB ID 4A8L), 1045D. Furthermore, the magnitude of the molecular charge of HEWL is two times smaller and of a positive nature than the charge of Glucose Isomerase that is of a negative nature.

The differences observed in the dipole moment, net molecular charge, and ratio between the number of positive and negative residues may explain the differences in field intensities reported to be required for the crystallization of HEWL. It may also explain the different behavior presented by GI and HEWL crystals grown in the E-field microbatch device. While GI crystals tended to grow closer to the bottom and sides of the well in the presence of an electrical field, HEWL crystals grown in the same crystallization device could be observed preferentially at the oil-growth solution interface. The later was also observed by Koizumi et al. [34] who reported that HEWL crystals tend to be located at the oil-growth solution interface when grown in the presence of an AC external electric field. There is no question that crystallization in the presence of an electrical field leads to higher quality crystals [19,37], as was also observed for the GI crystals analyzed that were grown in a DC external field.

The smaller number of GI crystals observed for the higher fields may be traced back to a decrease in the nucleation rate with the increase of the DC field. Smaller nucleation rates reflected in the reduced numbers of crystals have been reported for HEWL crystallized in the presence of constant electrical fields [18,19]. Recent crystallization studies in the presence of AC electric fields have demonstrated that it is possible to control the nucleation rate of HEWL [34], although electrical fields applied were two to three orders of magnitude higher.

Thermodynamic studies and the classical theory of nucleation [43,44] indicate that electric fields may increase or decrease the free energy required for nucleation. On the other hand, the nucleation barrier height to nucleation is proportional to surface tension, and therefore a small change in surface tension may lead to significant changes in the nucleation rate. Electric fields with a strength of 10^5 to 10^6 V/m were reported in the control of the nucleation of NaCl crystals and the crystallization of other small molecules in aqueous mediums [43]. These electric fields actually cause a decrease in the values of kinetic parameters and, in the case of small molecules, stimulate the nucleation rate for crystallization conditions in saturated solutions. Contrary to the results observed for the crystallization of small molecules in the presence of electric fields, a drastic decrease in the number of nuclei is observed for crystallization of biological macromolecules in the presence of constant electric fields [18,19,21,24]. In fact, it seems to be possible to control nucleation through AC electric fields [35].

All reports so far have concluded that the crystallization of macromolecules in the presence of electric fields leads to a decrease of nucleation sites, and crystals tend to be significantly larger in size and of better crystallographic quality. However, there are still several questions to be answered. It is therefore important to consider the nature and properties of the macromolecule to be crystallized in the presence of an electrical field. The knowledge of the molecular properties should allow for the optimization of the intensity and nature of the field to be applied, allowing the study of the effect of a single parameter in this multiparametric process that is the crystallization of macromolecules.

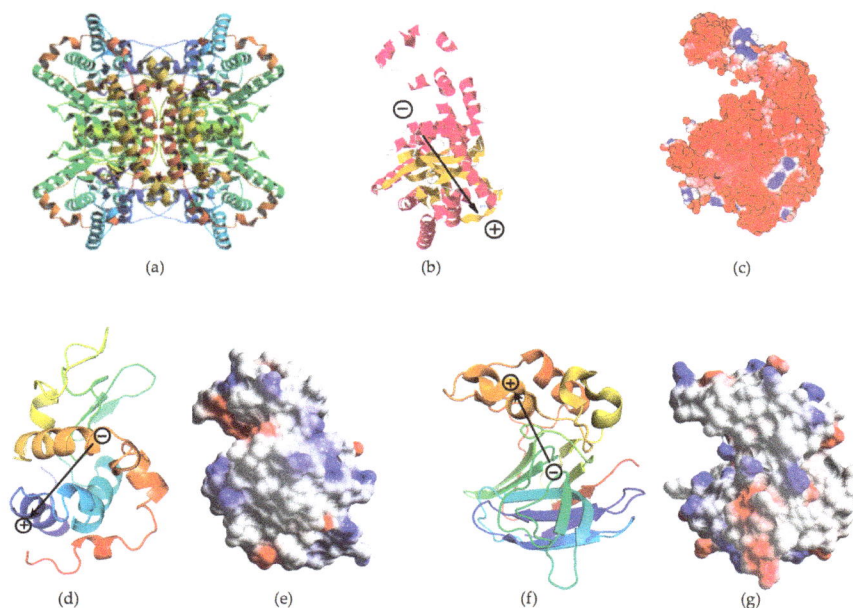

Figure 4. Top panel: Glucose Isomerase from *Streptomyces rubiginosus*, nearly 49% of the secondary structure is helical (192 residues in 20 helices) and only 10% is arranged in beta sheets (39 residues in 17 strands): (**a**) crystal structure (PDB ID 4A8L); (**b**) monomer showing the electric dipole moment vector (http://dipole.weizmann.ac.il); and (**c**) surface charge (Coot: http://lmb.bioch.ox.ac.uk/coot/). **Bottom panel**: Hen Egg White Lysozyme (HEWL) (PDB ID 193L) (**d**) monomer showing electric dipole moment and (**e**) surface charge; and Thaumatin (PDB ID 1RQW) (**f**) monomer showing electric dipole moment and (**g**) surface charge. Electric dipole moment vector was determined online through http://dipole.weizmann.ac.il and surface charge by Coot: http://lmb.bioch.ox.ac.uk/coot/.

Table 1. Electric dipole moment computed for several macromolecules commonly crystallized in the presence of electric fields, DC or AC.

	GI	HEWL	Thaumatin	HEHH	Ferritin
PDB ID	4A8L	193L	1RQW	1QYW	3F32
dipole moment (D)	1045	198	478	770	529
# of atoms	3187	1012	1604	2123	1411
# of residues	386	129	207	276	168
# of positive residues	46	16	24	29	19
# of negative residues	65	9	19	30	26
charge	−19	8	5	−1	−7

4. Materials and Methods

4.1. Electric Field Crystallization Device—Efield Microbatch

The *Efield microbatch device* was designed with the intent to take advantage of existing crystallization methods to study the effect of a DC external field on the crystallization process of macromolecules. The first design included a high voltage power supply; it was bulky and not easily accessible. A second design was developed focusing on the ease for temperature control and mobility.

The device is compact enough to fit into an incubator. The schematics and image of the device are shown in Figure 5.

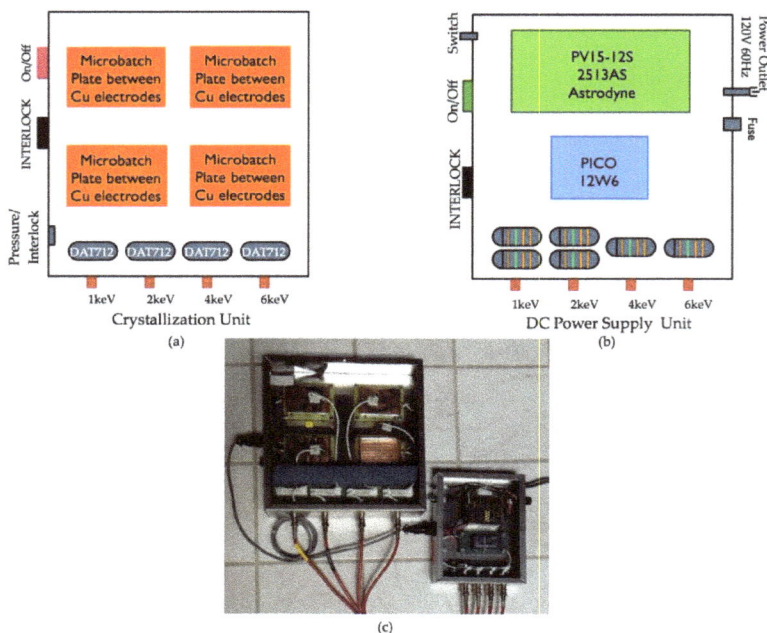

Figure 5. Crystal Growth Unit: External DC electric field set up. The instrument allows for four DC voltages 0, 1, 2, 4, and 6 KV to be applied to microbatch plates. Copper electrodes are 85 mm × 60 mm. The microbatch plates (Hampton Research) were 10 mm high. The bottom copper electrodes are fixed while the top electrodes are connected to a microbatch plate lid that is kept between crystallization trials. The device is interlocked during a crystallization experiment due to the high voltage being applied. Top view (**a**) schematics, (**b**) external DC electric field device with crystallization plates in place. Bottom view (**c**) the actual device showing the crystallization unit (left in the figure) and the power supply unit (right in the figure).

4.2. Electric Field Simulation

The electric field experiment was simulated to better visualize the field's effect on the protein droplets (Figure 6). Two programs were used in the simulation process: Autodesk Inventor 2013 and Ansoft Maxwell 3D version 12. Maxwell 3D is a simulation program that uses finite element analysis to solve electromagnetic fields on structures such as electric motors and other electromechanical devices. Inventor is a CAD program used for designing and drawing 3D objects. Inventor was used to draw the structures for the simulation, including the 72-well Terasaki plate with a lid, copper voltage plates, mineral oil droplets, and protein droplets. The structures were imported into Maxwell to perform the field simulations.

In Maxwell, the system was first defined as electrostatic, which set up the simulator for electric field calculations. The imported structures were assigned materials from Maxwell's materials library. The microbatch plate and lid were assigned polystyrene, the voltage plates assigned copper, and the background assigned air. The protein droplets were assigned seawater, a good approximation since the major components of the droplets are water and salt in high concentration. There was no mineral oil available in the Maxwell library, so the material was user-generated using the relative permittivity and conductivity of real mineral oil. Voltages were assigned to the copper electrodes, with the top electrode being a positive voltage between 1 kV and 6 kV and the bottom electrode being 0 V. The last step was assigning the level of solution accuracy. This value was set to achieve an error of less than 0.0001%, forcing the program to refine its finite mesh and recalculate until the high accuracy was met.

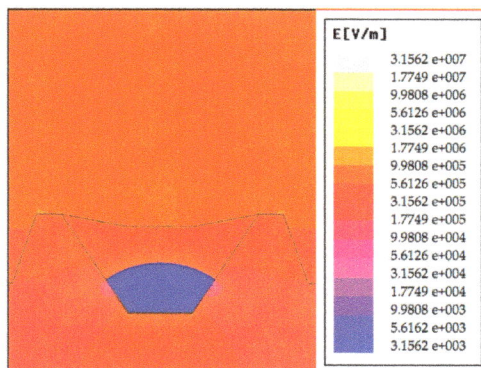

Figure 6. Electric field intensity simulation for the microbatch plates. A single well is shown for clarity reasons. The false color sequence indicates the intensity of the field.

4.3. Crystallization Conditions in Electric Fields

Glucose Isomerase from *Streptomyces rubiginosus* was crystallized in the microbatch method developed by Chayen [38,39]. Glucose Isomerase (HR7-102) purchased from Hampton Research was dialyzed versus 10 mM HEPES pH 7.0 and 1 mM magnesium chloride ($MnCl_2$) (Sigma-Aldrich, St. Louis, MO, USA). Concentration of the dialyzed solution was measured using a UV-VIS (Thermo Spectronic spectrophotometer; Thermo-Fisher, Waltham, MA, USA). Several crystallization conditions were screened to find an optimal precipitant, acidity, and protein concentration. Crystallization conditions were obtained by mixing 1 μL of 60 mg/mL Glucose Isomerase with 1 μL sodium citrate (Sigma Aldrich); drops were covered with 10 μL mineral oil (Sigma Aldrich). Each experiment consisted of five 72-well microbatch crystallization plates, HR3-081 (Hampton Research, Aliso Viejo, CA, USA). For each experiment, five plates were prepared and submitted to five different DC voltages: 0, 1, 2, 4, and 6 kV for a 48-h period. For each trial, protein and crystallization solutions were prepared fresh. Five individual experimental sets were carried out to verify reproducibility over a period of two month. All experiments were carried out at room temperature, 20 ± 1 °C. At the end of each experiment, the five microbatch plates were stored in an incubator at 18 °C. Each individual plate was analyzed with a visible light microscope equipped with polarizers (Olympus ZTE 70; Olympus Scientific, Waltham, MA, USA). Each well, 72 per crystallization plate, was photographed, and the number of crystals and their size were recorded (Figure 7). Crystal size was determined while pictures were being recorded directly from the microscope, and later dimensions were confirmed using the open source software ImageJ (https://imagej.net/ImageJ).

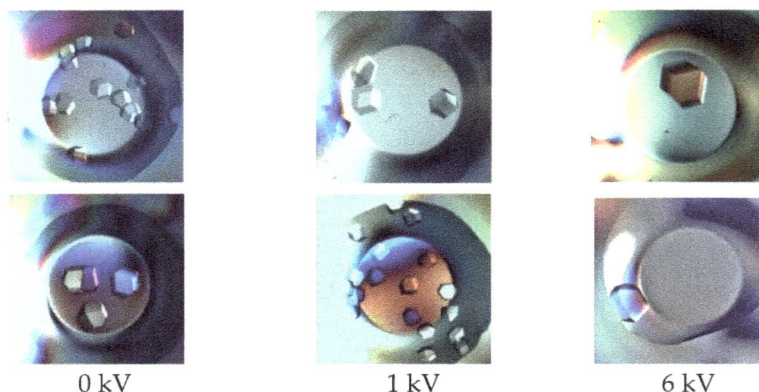

| 0 kV | 1 kV | 6 kV |

Figure 7. Efield microbatch crystals of Glucose Isomerase from *Streptomyces rubiginosus* in the presence of different DC electric field intensities. Each experiment consisted of five 72-well microbatch crystallization plates (Hampton Research). For the experiments, 1 μL of protein (Glucose Isomerase) and 1 μL of precipitant solution (sodium citrate) and 10 μL of mineral oil were placed in each well. The trials were exposed to five different DC voltages: 0, 1, 2, 4, and 6 kV for a 48-h period. The top and bottom panels are typical results from experiments realized on separate dates.

4.4. Crystal Quality Assessment

For each condition, five Glucose Isomerase crystals of similar dimensions were randomly selected for each experiment. Crystals were mounted on MicroRT™ (MiTeGen LLC) capillary system. Then, 5 μL of mother liquor was added to the transparent polymer capillaries to prevent dehydration. X-ray diffraction data was collected at the X6A beam line at the National Synchrotron Light Source; the wavelength was kept constant at 0.799 Å and crystal rotation per frame 0.5°. Exposure times varied between 1 s and 10 s, depending on the electric field applied. For each crystal, 40 images and a total of 20 degrees were recorded, however, to minimize possible effects of radiation damage, only the first 10 frames were used to assess the crystal quality. All diffraction images were analyzed with the HKL2000 software package [45]. Parameters such as mosaicity, resolution, and signal-to-noise ratio (I/σI) were recorded and reported in this paper. For each experiment and for each individual crystallization plate, five crystals were exposed to X-rays for statistical analyses; in total, 75 crystals were processed for statistics. The average mosaicity and resolution recorded for each of the applied voltages were recorded (Figure 3).

Acknowledgments: We thank Dennis Poshka for his assistance in the design and construction of the external DC electric field device. Evgeniya Rubin and Christopher Owen thank the support received from the DOE Summer Undergraduate Laboratory Internship (SULI) program. The X6A beam line is funded by the National Institute of General Medical Sciences under agreement GM-0080. The National Synchrotron Light Source, Brookhaven National Laboratory was supported by the US Department of Energy under contract #DE-AC02-98CH10886. The LSBR is funded by the Department of Energy (DOE) BER program and the NIH NIGMS grant #.

Author Contributions: Vivian Stojanoff conceived and designed the experiments; Evgeniya Rubin and Vivian Stojanoff performed the experiments; Christopher Owen build and tested the DC external electric field device under Vivian Stojanoff guidance; Evgeniya Rubin, Christopher Owen and Vivian Stojanoff analyzed the data; Vivian Stojanoff wrote the paper.

Conflicts of Interest: The authors declare no conflict of interest. The founding sponsors had no role in the design of the study; in the collection, analyses, or interpretation of data; in the writing of the manuscript, and in the decision to publish the results.

References

1. McPherson, A. *Crystallization of Biological Macromolecules*; Cold Spring Harbor Laboratory Press: New York, NY, USA, 1999; 402p.

2. Chayen, N.E.; Helliwell, J.R., Jr.; Snell, E.H. Macromolecular Crystallization and Crystal Perfection. In *IUCr Monographs on Crystallography 24*; Oxford University Press: Oxford, UK, 2010; 21p., ISBN 978-0-19-921325-2.

3. Allegre, C.J.; Provost, A.; Jaupart, C. Oscillatory zoning: A pathological case of crystal growth. *Nature* **1981**, *294*, 223–228. [CrossRef]

4. Vergara, A.; Lorber, B.; Zagari, A.; Giege, R. Physical aspects of protein crystal growth investigated with the Advanced Protein Crystallization Facility in reduced-gravity environments. *Acta Crystallogr. D* **2003**, *59*, 2–15. [CrossRef] [PubMed]

5. McPherson, A.; DeLuca, L.J. Microgravity protein crystallization. *NPJ Microgravity* **2015**, *1*, 15010. [CrossRef]

6. Petsev, D.N.; Thomas, B.R.; Yau, S.; Vekilov, P.G. Interactions and aggregation of apoferritin molecules in solution: Effects of added electrolytes. *Biophys. J.* **2000**, *78*, 2060–2069. [CrossRef]

7. Garcia-Ruiz, J.M.; Moreno, A.; Viedma, C.; Coil, M. Crystal quality of lysozyme single crystas grown by the gel acupuncture method. *Mater. Res. Bull.* **1993**, *28*, 541–546. [CrossRef]

8. Garcia-Ruiz, J.M. Counterdiffusion Methods for Macromolecular Crystallization. *Methods Enzimol.* **2003**, *368*, 130–154. [CrossRef]

9. Rummel, G.; Hardmeyer, A.; Widmer, C.; Chiu, M.L.; Nollert, P.; Locher, K.P.; Pedruzzi, I.; Landau, E.M.; Rosenbusch, J.P. Lipidic Cubic Phases: New Matrices for the three-dimensional crystallization of membrane proteins. *J. Struct. Biol.* **1998**, *121*, 82–91. [CrossRef] [PubMed]

10. Caffrey, M. A comprehensive review of the lipid cubic phase or in meso method for crystallizing membrane and soluble porteins and complexes. *Acta Crystallogr. F* **2015**, *71*, 3–18. [CrossRef] [PubMed]

11. Chayen, N.E. A novel technique for containerless protein crystallization. *Protein Eng. Des. Sel.* **1996**, *9*, 927–929. [CrossRef]

12. Govada, L.; Leese, H.S.; Saridakis, E.; Kassen, S.; Chain, B.; Khurshid, S.; Menzel, R.; Hu, S.; Shaffer, M.S.P.; Chayen, N.E. Exploring Carbon Nanomaterial Diversity for Nucleation of Protein Crystals. *Sci. Rep.* **2016**, *6*, 20053. [CrossRef] [PubMed]

13. Bergfors, T. Screening and optimization methods for nonautomated crystallization laboratories. *Methods Mol. Biol.* **2007**, *363*, 131–151. [PubMed]

14. Bergfords, T. Seeds to crystals. *J. Struct. Biol.* **2003**, *142*, 66–76. [CrossRef]

15. Wakayama, N.I. Effects of a Strong Magnetic Field on Protein Crystal Growth. *Cryst. Growth Des.* **2003**, *3*, 17–24. [CrossRef]

16. Surade, S.; Ochi, T.; Nietlispach, D.; Chirgadze, D.; Moreno, A. Investigations into Protein Crystallization in the Presence of a Strong Magnetic Field. *Cryst. Growth Des.* **2010**, *10*, 691–699. [CrossRef]

17. Moreno, A.; Yokaichiya, F.; Dimasi, E.; Stojanoff, V. Growth and Characterization of high-quality protein crystals for X-ray crystallography. *Ann. N. Y. Acad. Sci.* **2009**, *1161*, 429–436. [CrossRef] [PubMed]

18. Hammadi, Z.; Veesler, S. New approaches on crystallization under electric fields. *Prog. Biophys. Mol. Biol.* **2009**, *101*, 38–44. [CrossRef] [PubMed]

19. Pareja-Rivera, C.; Cuéllar-Cruz, M.; Esturau-Escofet, N.; Demitri, N.; Polentarutti, M.; Stojanoff, V.; Moreno, A. Recent Advances in the Understanding of the Influence of Electric and Magnetic Fields on Protein Crystal Growth. *Cryst. Growth Des.* **2017**, *17*, 135–145. [CrossRef]

20. Chin, C.C.; Dence, J.B.; Warren, J.C. Crystallization of human placental estradiol 17beta-dehydrogenase. A new method for crystallizing labile enzymes. *J. Biol. Chem.* **1976**, *251*, 3700–3705. [PubMed]

21. Taleb, M.; Didierjean, C.; Jelsch, C.; Mangeot, J.P.; Capelle, B.; Aubry, A. Crystallization of proteins under an external electric field. *J. Cryst. Growth* **1999**, *200*, 575–582. [CrossRef]

22. Taleb, M.; Didierjean, C.; Jelsch, C.; Mangeat, J.P.; Aubry, A. Equilibrium kinetics of lysozyme crystallization under an external electric field. *J. Cryst. Growth* **2001**, *232*, 250–255. [CrossRef]

23. Ries-Kautt, M.; Ducruix, A. From solution to crystals with a physico-chemical aspect. In *Crystallization of Nucleic Acids and Proteins: A Practical Approach*, 2nd ed.; Ducruix, A., Giégé, R., Eds.; Oxford University Press: Oxford, UK, 1999; pp. 269–312.

24. Nanev, C.N.; Penkova, A. Nucleation of lysozyme crystals under external electric, ultrasonic fields. *J. Cryst. Growth* **2001**, *232*, 285–293. [CrossRef]

25. Nanev, C.N.; Penkova, A. Nucleation and growth of lysozyme crystals under external electric field. *Colloids Surf. A Physicochem. Eng. Asp.* **2002**, *209*, 139–145. [CrossRef]

26. Garetz, B.A.; Matic, J.; Myerson, A.S. Polarization Switching of Crystal Structure in the Nonphotochemical Light-Induced Nucleation of Supersaturated Aqueous Glycine Solutions. *Phys. Rev. Lett.* **2002**, *89*, 175501. [CrossRef] [PubMed]

27. Penkova, A.; Gliko, O.; Dimitrov, I.L.; Hodjaoglu, F.V.; Nanev, C.; Vekilov, P.G. Enhancement and suppression of protein crystal nucleation due to electrically driven convection. *J. Cryst. Growth* **2005**, *275*, e1527–e1532. [CrossRef]

28. Mirkin, N.; Frontana-Uribe, B.A.; Rodriguez-Romero, A.; Hernandez-Santoyo, A.; Moreno, A. The influence of an internal electric field upon protein crystallization using the gel-acupuncture method. *Acta Crystallogr.* **2003**, *D59*, 1533–1538. [CrossRef]

29. Moreno, A.; Sazaki, G. The use of a new ad hoc growth cell with parallel electrodes for the nucleation control of lysozyme. *J. Cryst. Growth* **2004**, *264*, 438–444. [CrossRef]

30. Nieto-Mendoza, E.; Frontana-Uribe, B.A.; Sazaki, G.; Moreno, A. Investigations on electromigration phenomena for protein crystallization using crystal growth cells with multiple electrodes: Effect of the potential control. *J. Cryst. Growth* **2005**, *275*, e1437–e1446. [CrossRef]

31. Gil-Alvaradejo, G.; Ruiz-Arellano, R.R.; Owen, C.; Rodriguez-Romero, A.; Rudino-Pinera, E.; Antwi, M.K.; Stojanoff, V.; Moreno, A. Novel Protein Crystal Growth Electrochemical Cell for Applications in X-ray Diffraction and Atomic Force Microscopy. *Cryst. Growth Des.* **2011**, *11*, 3917–3922. [CrossRef]

32. Flores-Hernandez, E.; Stojanoff, V.; Arreguin-Espinosa, R.; Moreno, A.; Sanchez-Puig, N. An electrically assisted device for protein crystallization in a vapor-diffusion setup. *J. Appl. Crystallogr.* **2013**, *46*, 832–834. [CrossRef] [PubMed]

33. Martinez-Caballero, S.; Cuellar-Cruz, M.; Demitri, N.; Polentarutti, M.; Rodriguez-Romero, A.; Moreno, A. Glucose Isomerase Polymorphs Obtained Using an Ad Hoc Protein Crystallization Temperature Device and a Growth Cell Applying an Electric Field. *Cryst. Growth Des.* **2016**, *16*, 1679–1686. [CrossRef]

34. Koizumi, H.; Fujiwara, K.; Uda, S. Role of electric double layer in controlling the nucleation rate for tetragonal hen egg white lysozyme crystals by application of an external electric field. *Cryst. Growth Des.* **2010**, *10*, 2591–2595. [CrossRef]

35. Koizumi, H.; Uda, S.; Fujiwara, K.; Tachibana, M.; Kojima, K.; Nozawa, J. Improvement of crystal quality for tetragonal hen egg white lysozyme crystals under application of an external alternating current electric field. *J. Appl. Crystallogr.* **2013**, *46*, 25–29. [CrossRef]

36. Koizumi, H.; Uda, S.; Fujiwara, K.; Tachibana, M.; Kojima, K.; Nozawa, J. Crystallization of high-quality protein crystals using an external electric field. *J. Appl. Crystallogr.* **2015**, *48*, 1507–1513. [CrossRef]

37. Koizumi, H.; Uda, S.; Fujiwara, K.; Tachibana, M.; Kojima, K.; Nozawa, J. Technique for High-Quality Protein Crystal Growth by Control of Subgrain Formation under an External Electric Field. *Crystals* **2016**, *6*, 95. [CrossRef]

38. Chayen, N.E.; Shaw Stewart, P.D.; Blow, D.M. Microbatch crystallization under oil—A new technique allowing many small-volume crystallization trials. *J. Cryst. Growth* **1992**, *122*, 176–180. [CrossRef]

39. Chayen, N.E.; Shaw Stewart, P.D.; Maeder, D.L.; Blow, D.M. An automated system for micro-batch protein crystallization and screening. *J. Appl. Cryst.* **1990**, *23*, 297–302. [CrossRef]

40. Stojanoff, V. Crystal quality a quest for structural proteomics. In *Synchrotron Radiation and Structural Proteomics*; Riekel, C., Ed.; Pan Stanford Publishing: Temasek Boulevard, Singapore, 2011; pp. 383–407.

41. Felder, C.F.; Prilusky, J.; Silman, I.; Sussman, J.L. A server and database for dipole moments of proteins. *Nucleic Acids Res.* **2007**, *35*, W512–W521. [CrossRef] [PubMed]

42. Emsley, P.; Lohkamp, B.; Scott, W.G.; Cowtand, K. Features and development of Coot. *Acta Crystallogr.* **2010**, *D66*, 486–501. [CrossRef]

43. Saban, K.; Thomas, J.; Varughese, P.A.; Varghese, G. Thermodynamics of Crystal Nucleation in an External Electric Field. *Cryst. Res. Technol.* **2002**, *37*, 1188–1199. [CrossRef]

44. Warshavsky, V.B.; Bykov, T.V.; Zeng, X.C. Effects of external electric field on the interfacial properties of weakly dipolar fluid. *J. Chem. Phys.* **2001**, *114*, 504–512. [CrossRef]
45. Otwinowski, Z.; Minor, W. Processing of X-ray diffraction data collected in oscillation mode. *Method Enzymol.* **1997**, *276*, 307–362.

crystals

MDPI

Article

Crystal Growth of High-Quality Protein Crystals under the Presence of an Alternant Electric Field in Pulse-Wave Mode, and a Strong Magnetic Field with Radio Frequency Pulses Characterized by X-ray Diffraction

Adela Rodríguez-Romero *, Nuria Esturau-Escofet, Carina Pareja-Rivera and Abel Moreno *

Instituto de Química, Universidad Nacional Autónoma de México, Av. Universidad 3000, UNAM, Ciudad de México 04510, Mexico; esturau_nuria@comunidad.unam.mx (N.E.-E.); carina.pareja@gmail.com (C.P.-R.)
* Correspondence: adela@unam.mx (A.R.-R.); carcamo@unam.mx (A.M.);
  Tel.: +52-55-5622-4568 (A.R.-R.); +52-55-5622-4467 (A.M.)

Academic Editor: Abel Moreno
Received: 29 April 2017; Accepted: 15 June 2017; Published: 19 June 2017

Abstract: The first part of this research was devoted to investigating the effect of alternate current (AC) using four different types of wave modes (pulse-wave) at 2 Hz on the crystal growth of lysozyme in solution. The best results, in terms of size and crystal quality, were obtained when protein crystals were grown under the influence of electric fields in a very specific wave mode ("breathing" wave), giving the highest resolution up to 1.34 Å in X-ray diffraction analysis compared with controls and with those crystals grown in gel. In the second part, we evaluated the effect of a strong magnetic field of 16.5 Tesla combined with radiofrequency pulses of 0.43 μs on the crystal growth in gels of tetragonal hen egg white (HEW) lysozyme. The lysozyme crystals grown, both in solution applying breathing-wave and in gel under the influence of this strong magnetic field with pulses of radio frequencies, produced the larger-in-size crystals and the highest resolution structures. Data processing and refinement statistics are very good in terms of the resolution, mosaicity and Wilson B factor obtained for each crystal. Besides, electron density maps show well-defined and distinctly separated atoms at several selected tryptophan residues for the crystal grown using the "breathing wave pulses".

Keywords: lysozyme; crystal growth in solution; gel-growth; electric fields; magnetic fields; pulse-wave

1. Introduction

The new trend in protein crystallography has been focused on finding new strategies to obtain crystals of different sizes for different bio-structural projects at very high resolution via X-ray diffraction (powder and single-crystal), neutron diffraction, and recently using X-ray free electron lasers (XFEL) methods. Particularly, neutron crystallography is an emerging research area in which we can follow the details of precise mechanisms of chemical interactions and the new strategies for solving the structure of macromolecular complexes. This is a powerful complementary technique for X-ray structural studies that explicitly determines the location and orientation of deuterium atoms in proteins [1–5]. However, this technique requires larger protein crystals than those used for conventional X-ray crystallography due to the low flux of neutron beams, though longer data collection (time varying from days to weeks) are also needed. That is why just a few 3D structures using neutron diffraction have been deposited in the Protein Data Bank (PDB). Despite these apparent disadvantages, there have been recent significant advances, not only in the beam lines and the detectors technology, but also in the sample preparation (perdeuteration), though the second via to obtain the 3D structure from nanocrystals is the free-electron lasers to perform X-ray crystallography of macromolecular complexes [6–12]. Additionally, magnetic fields [13–18], electric fields [16,19,20] or a combination of both [21] have proven to be efficient in obtaining bigger and higher-quality protein single crystals for X-ray crystallography. The use of alternate currents (AC), has also been explored in protein crystallization focused on the size of protein and on the quality of the crystals, using X-ray topography [20]. There are also other novel strategies and ad hoc devices that have been used in electric (DC & AC) [22–30] and ultrasonic fields [31] applied to protein crystallization techniques. The pioneering contributions about the use of electric fields applied to protein crystallization were proposed by Professor Christo Nanev and his group more than a decade ago [31–33] followed by a research on the kinetics of the process by Professor Aubry and his group [34,35]. In this line of research, we found additional related ideas using pulses such as femto-lasers instead of electric fields for controlling the nucleation [36]. The use of lasers permitted to observe the area where the laser struck produced the formation of just crystalline nuclei (this can be explained by the formation of small fragments of protein serving as seeds for growing nucleation centers produced by the focalized laser radiation) [37]. An alternative method would be the use of agarose for growing high-quality crystals. These crystals would be suitable not only for high resolution X-ray crystallography, but also for soaking ligands analogs in enzymes related to protein biosynthesis. This gel-growth technique or the counter-diffusion in capillary tubes method can also be combined with different devices using electromagnetic fields [18,21]. In the last decade, agarose was the most popular gel for protein crystallization, however other gels were also used in protein crystallization and in crystal growth of inorganic and organic compounds [38–44].

In this contribution, we evaluated the effect of four types of pulse waves when applying an AC current of 2 Hz to the crystal growth of lysozyme in solution. The best crystals, in terms of size and quality, were grown in solution with the pulse wave called "breathing wave". Additionally, a strong magnetic field of 16.5 Tesla combined with radiofrequency pulses of 0.43 μs were applied also on the crystal growth of lysozyme in gels. The lysozyme crystals grown both in solution (with AC in breathing wave mode) and in gel under the influence of this strong magnetic field, produced large-sized crystals and high-resolution crystallographic structures. Data processing and refinement statistics are very good in terms of the crystal quality, as well as the electron density maps of the best crystals.

2. Results and Discussion

The crystal growth cell (as shown in Figure 1) used in these experiments for applying electric field (AC) in four types of wave modes was based on the batch method for proteins crystallization that we had previously published using DC, though with some modifications [16]. In the current experiments, we have tried to investigate the effect of alternate current (AC) in dense-disperse wave mode (usually called pulse-wave) at 2 Hz in four different modes on the crystal growth of lysozyme in solution.

The device used for these alternating current experiments was a KWD-808 (Jiangsu, China) I Multi-purpose health device (usually recommended for acupuncture treatment). This device allows to apply five types of waves in alternant current (AC) at different frequencies (f): the first pulse wave corresponds to the continues-wave, it contains repeated pulses ranging from 0.5 Hz up to 100 Hz, the maximum output amplitude pulse is from 40 V ± 10 V. This type of continues pulse related to AC has been used in several other publications elsewhere.

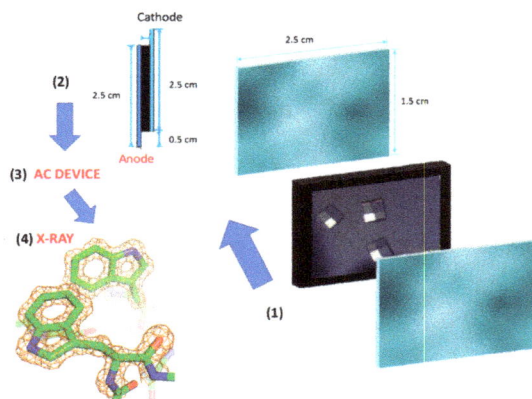

Figure 1. The growth cell (**1**) and the setup used along these experiments for the electric field approach (**2,3**) up to the 3D structure by X-ray crystallography (**4**). The electrodes are made of Indium Tin Oxide (ITO) glass. The electron density maps (2Fo-Fc) of the best crystals at a contour level of 1σ is shown after X-ray crystallographic analysis.

In our case, Figure 2 shows the effect of alternate current (AC) by using four types of pulse waves: Type I (scheme shown below Figure 2A), this pulse is called dense- and scattered-wave; the second type II (scheme shown below Figure 2B) is called intermittent-wave; the third type III (scheme shown below Figure 2C) is called ripple-wave, which reminds us of the waves on the surface of water, especially caused by an object dropping into it; the fourth type IV (scheme shown below Figure 2D) is called breathing pulse, which implies that the wave first rises its amplitude and then falls to the ground level. From our results, crystals obtained under pulse wave type I were very tiny, which made these crystals hard to characterize using our rotating anode X-ray generator. Nonetheless, the left three types of pulse waves (Figure 2B–D) produced suitable crystals for X-ray diffraction. When applying these four pulse waves, each of them presented a variation on the number and on the size of the crystals along the cell. The best crystals (in terms of size and number) were obtained when wave pulse type IV (called "breathing wave") produced the smallest number of crystals, but larger in size (Figure 2D).

We statistically measured the number of crystals attached to the type of electrode (cathode or anode) versus the pulse wave as shown in Figure 3. It was remarkable to see that the type IV of pulse wave (Figure 2D called "breathing wave") produced the smallest number of crystals, though statistically larger in size as shown in Figure 4. The size of these crystals was very regular compared to the other types of waves.

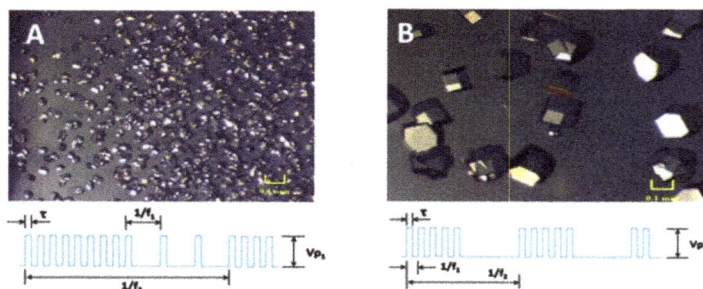

Figure 2. *Cont.*

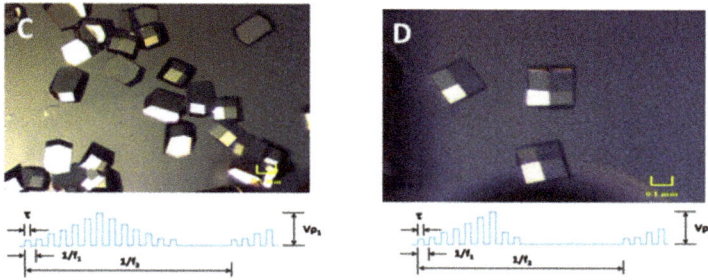

Figure 2. Four types of pulse waves were tested (**A–D**) as shown underneath each picture. The last pulse wave, where better crystals are obtained, is called breathing pulse wave.

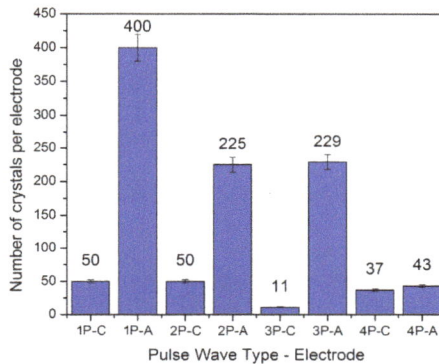

Figure 3. The pulse (P) wave number corresponds to the type of wave-pulse (I to IV mentioned in the text). This shows the number of crystals attached to the anode (A) or to the cathode (C).

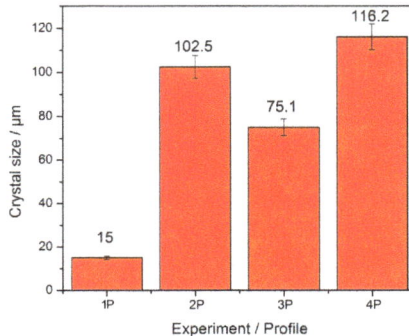

Figure 4. The crystals size is plotted versus type of pulse (which corresponds to type I to IV in the text) on the surface of the electrode, where the smallest number of crystals were observed.

In Figure 4, we see that the standard deviation (error bar) of pulse wave 4P is higher than the one of pulse wave 1P. This is due to the bigger size of the crystals though fewer in number, which pulse wave 4P produced. Therefore, in terms of the crystals number for pulses 1P and 4P, we must refer to Figure 3. Here we see that pulse wave 4P, called breathing wave, had fewer number of crystals distributed on the surface of each of the electrodes (anode and cathode), whereas pulse wave 1P had a greater number of crystals on the electrodes. However, 1P crystals were a lot smaller and unevenly distributed (400 on the anode and 50 on the cathode). The statistical measurement of the size of crystals (Figure 4) showed a more homogeneous size distribution of crystals per square centimeter in pulse wave 4P than in 1P, even though the error bar in 1P is smaller than in the rest. In this Figure 4, we see that pulse wave 2P, 3P and 4P also have different size error bars. Again, this is due to the crystals size variations (from 75 to 116 microns) in these pulse waves, whereas the size variation in pulse wave 1P is only around 15 microns. There is one order of magnitude of difference for each of the comparable sizes (1P compared to 2P, 3P and 4P). The fact that the error bar is lower for 1P than for the rest is due to the tiny size of the crystals that 1P produced, regardless of their number. The only difference between the crystals in 2P, 3P and 4P was the refinement statistics and the crystal quality (see Table 1).

We applied a statistical analysis by using the Student's test. In this test, we compared the crystal sizes of 1P with the ones obtained using pulse waves (2P, 3P and 4P). For each of these comparisons we obtained different values of variance: 1P and 2P (variance: 3.0909×10^{-6} and 1.5832×10^{-4}); 1P and 3P (variance: 3.0909×10^{-6} and 4.0621×10^{-5}); 1P and 4P (variance: 3.0909×10^{-6} and 2.1781×10^{-4}). This means that the samples are independent. The p-value for all samples was less than 0.05, which means that there is a statistically significant difference between 1P and 2P or 1P and 3P or 1P and 4P. These results were already discussed in the previous paragraph. The mean size values are also different when comparing 1P (0.015 mm) with 2P (0.103 mm), 3P (0.075 mm) or 4P (0.116 mm). The degrees of freedom 34 (2P), 20 (3P) and 15 (4P) are related to the number of measured crystals with a confidence of 95%.

Recent publications have revealed that the use of related types of e-crystallization growth cells applied to proteins by using direct current (DC) made crystals grow better oriented to the cathode (when the protein molecule was positively charged) compared to those crystals grown on the anode (when the protein molecule was negatively charged) showing also high crystal quality [16–20]. Along the crystal growth process, four different regimes are usually obtained: (1) induction/equilibration; (2) transient nucleation; (3) steady-state nucleation & crystal growth; and (4) depletion [45]. However, from the present results we observed that AC in a pulse-wave mode produced better results when a specific pulse wave was used ("breathing-wave").

Nonetheless, it has been observed that a homogeneous magnetic field, when applied for a long time, reduces the gravity forces on the solution through the action of the magnetic force. This has a positive effect on the crystal growth [16,18]. The convection is practically nullified, generating a situation like the one found under the conditions of microgravity. Based on this assumption, we decided to grow lysozyme crystals in gels, under the presence of a strong magnetic field applying pulses of radio frequency at different angles.

The experimental setup for these results using a strong magnetic field on the crystallization of lysozyme is shown in Figure 5. For this case, we used a magnetic field of 16.5 Tesla combined with crystal growth in gels and cycles of rectangular radiofrequency pulses of 0.43 µs followed by a delay time of 4.3 s. This is the first time that short radio frequency pulses have been applied for growing crystals in gels and inside the NMR spectrometer of 700 MHz (16.5 Tesla). The lysozyme crystals were larger and of a better quality than those of the controls (see Table 1), improving even the results obtained in our most recent publication [16]. It is important to remark that these lysozyme crystals not only showed the typical crystal orientation along the c-axis, but they also were larger compared with the best controls of crystals grown in solution and in gels. Compared to the results of reference [18], where only seven Tesla were applied, the size of the crystals was comparable to the direction of the "C" axis, but the crystals in our experiments were much more elongated.

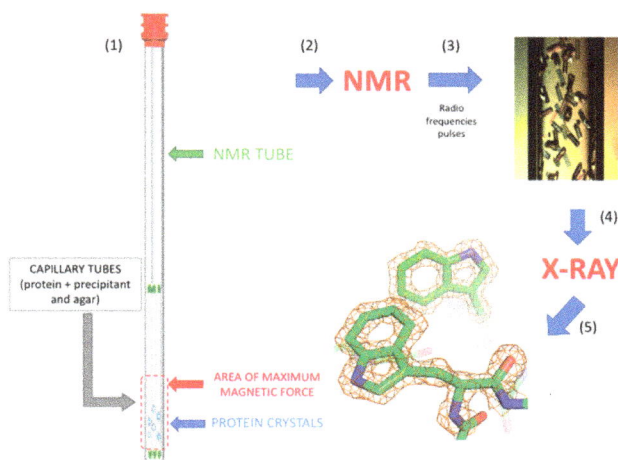

Figure 5. Experimental setup for the crystal growth in gels under the influence of magnetic field of 16.5 Tesla applying radio frequency pulses. The numbers represent the sequence of experiments (**1–4**) until obtaining the 3D structure (**5**). The capillary tube between (**3**) and (**4**) is 1 mm in diameter. The 3D structure shown after (**5**) is the electron density maps (2Fo-Fc) of the best crystals at a contour level of 1σ. This selected area comprises tryptophan residues (Trp 28 and 108) showing the crystal quality.

We collected full datasets of the best crystals, obtained by testing electric and magnetic fields, to get their 3D structures, so they could be compared with control crystals. Analysis of data collection and refinement statistics clearly showed that the crystal obtained when applying the "breathing pulse wave" improved the internal order of crystals and thus gave the better statistics in terms of resolution, mosaicity and B factor (Wilson Plot) (Table 1) and a more precise three-dimensional structure at 1.34 Å resolution. As shown in Table 1, crystal structures show slightly higher B factor value for those crystals grown in gels and NMR, indicating that they exhibit higher flexibility compared to those grown using pulses. Conversely, mosaicity values for the crystal grown using magnetic field was slightly lower (Lys-Gel 0.45°) than those obtained for the crystals grown using electric fields (0.81°, 0.55° and 0.50° for Lys2HzP2, Lys2HzP3 and Lys2HzP4, respectively).

Figure 6A–D shows the electron density map of the best crystals obtained from these experiments. The selected area was a relatively interior region that contains tryptophan residues 28 and 108 to show the crystal quality in terms of the definition of the indole rings. The crystals grown in gel, used as controls, also showed a good crystal quality as seen in the statistics table of the X-ray analysis (Table 1). However, the crystal obtained when applying the breathing pulse wave (see the electron density map on Figure 6B) corresponding to the Lys2HzP4 data in Table 1, presented the highest crystal quality compared to all analyzed crystals. These crystals diffracted up to 1.34 Å resolution. Nonetheless, several residues exposed to the solvent, such as Lys97 and Arg125, did show a dynamically disordered end of the side chain and not interpretable electron density and several showed double conformations (Arg45, Ser85, Ser86) in all the structures. This is the first time that this type of experiment shows a practical method for obtaining high-quality crystals using different types of pulse waves in solution.

Table 1. Data collection and refinement statistics of the best crystals obtained by using these methods compared with controls.

	Control (Solution)	Control (gel)	Lys-Gel (16.5 Tesla)	Lys2HzP2 (Figure 2B)	Lys2HzP3 (Figure 2C)	Lys2HzP4 (Figure 2D)
Data collection and processing statistics						
Wavelength (Å)	1.54	1.54	1.54	1.54	1.54	1.54
Space group	$P4_32_12$	$P4_32_12$	$P4_32_12$	$P4_32_12$	$P4_32_12$	$P4_32_12$
Unit-cell parameters (Å, °)	a = 77.43 b = 77.43 c = 37.38	a = 78.27 b = 78.27 c = 37.21	a = 78.27 b = 78.27 c = 37.21	a = 77.36 b = 77.36 c = 37.85	a = 78.44 b = 78.44 c = 37.08	a = 78.31 b = 78.31 c = 37.44
Reflections (unique)	10,721	19,481	22,482	18,167	22,596	25,225
Resolution limits (Å)	24.5–1.82	39.16–1.49	39.22–1.39	38.79–1.49	39.25–1.39	39.15–1.34
Completeness (%)	99.9 (100.0)	99.40 (90.0)	99.40 (86.0)	99.9 (99.9)	99.9 (99.2)	100 (99.7)
R_{merge} (%)	3.1 (12)	4.1 (19)	4.3 (19)	4.4 (39)	3.5 (20)	3.4 (34)
$I/\sigma(I)$	77 (4.5)	76.8 (3.6)	79 (3.1)	68 (3.9)	70 (5.2)	72 (4.2)
Average multiplicity	13.7 (4.0)	15.1 (7.3)	15.1 (7.3)	14.6 (8.1)	3.3 (3.1)	14 (4.1)
B factor from Wilson plot ($Å^2$)	18.7	18.2	15.1	17.2	14.6	14.1
Refinement statistics						
R/R_{free}	0.17/0.21	0.181/0.21	0.178/0.217	0.172/0.221	0.177/0.22	0.169/0.193
No. of protein atoms	1032	1002	1032	1197	1034	1199
No. of water molecules	129	127	163	190	176	181
No. of other atoms		3	12	6	12	19
R.m.s. deviations						
Bonds lengths (Å)	0.017	0.018	0.018	0.021	0.017	0.014
Bonds angles (°)	1.744	1.991	1.890	1.960	1.59	1.401
Average B factor ($Å^2$)	18.70	20.00	17.43	21.28	16.60	15.70

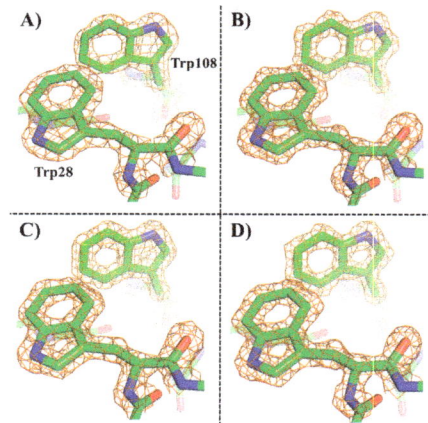

Figure 6. Comparison of electron density maps (2Fo-Fc) of the best crystals at a contour level of 1σ. This selected area is for side chains of tryptophan residues (Trp 28 and 108) showing the crystal quality: (**A**) control crystals in solution; (**B**) breathing pulse wave (pulse number 4, see the text); (**C**) crystal growth in gel (control); and (**D**) crystals grown in gel and magnetic field with radio frequencies pulses.

The analyses of the crystal structure by X-ray diffraction showed the highest crystal quality for those crystals grown using the pulse wave called breathing pulse type (see Figure 6B and X-ray data on Table 1 column Lys2HzP4. Not only were the crystals bigger, they were also remarkably reduced in number. These crystals showed the higher resolution limits (1.34 Å), lower Wilson B factors values, (15.7 $Å^2$) and low mosaicity values (0.50°). Unfortunately, there are not at the moment similar experiments in the literature to be compared with our results on protein crystallization, neither in electric fields (AC) using pulse waves nor using magnetic fields of 16.5 Tesla with radio frequencies pulses. There is just another crystallographic approach, in the same line, that uses a strong magnetic field of 10 Tesla applied to the crystallization of orthorhombic lysozyme with space group $P2_12_12_1$ [46]. This orthorhombic lysozyme structure was refined to a resolution of 1.13 Å and an R factor of 17%. The crystallographic analysis showed that few residues were shifted (Arg68, Arg73, Arg128) resulting in significant structural fluctuations, which can have large effects on the crystallization process and properties of lysozyme. What the authors claimed was that the strong magnetic field of 10 Tesla contributed to the stabilization of the dihedral angles. The significant difference between our results applied to tetragonal lysozyme with a space group $P4_32_12$ and those of Saijo et al. [46] was that they obtained a Mean B factor of 17.8 Å^2 after applying 10 Tesla (19.5 Å^2 for 0 Tesla), whereas we obtained a Mean B factor of 17.43 Å^2 after applying 16.5 Tesla (20 Å^2 for 0 Tesla). Nonetheless, the best Mean B factor in our work was the one obtained for the breathing pulse (4P) crystals (15.7 Å^2). The resolution in Saijo et al. [46] was improved from 1.33 to 1.13 Å (R factor of 17%), whereas in our case, the resolution factor was improved from 1.82 to 1.39 Å (23%). The mosaicity values for crystals grown using magnetic field of 10 Tesla was better than the lysozyme controls. In our case, we obtained similar results, the mosaicity was better when using 16.5 Tesla (0.45°) than those values obtained for the control (0.81°). From the structural point of view at 10 Tesla, Saijo et al. [46] observed the displacement of the charged side chains of Arg68 and Arg73 in the flexible loop and of Arg128 at the C-terminus. In our crystallographic analysis for tetragonal lysozyme grown applying 16.5 Tesla in gels with radio frequencies pulses, these displacements were almost the same although both structures have two different crystallographic space groups. However, in our case several residues exposed to the solvent, such as Lys97, Arg125, Arg128 and even the benzene ring of Trp62 showed a dynamically disordered end of the side chain and not interpretable electron density. Besides, several side chains showed double conformations (Arg45, Ser85, Ser86) in all the structures. Overall, both the magnetic field of 16.5 Tesla and the use of AC using the pulse wave called breathing pulse (4P), improved the crystal quality of tetragonal lysozyme. In both cases, the diffracted intensities increased significantly with these two approaches, leading to a higher resolution and better 3D crystallographic structures.

3. Materials and Methods

3.1. Gel Preparation

Agarose gel 1.0 % (*w/v*) stock solution of low melting point agarose (T_{gel} = 297–298 K, Hampton Research Cod. HR8-092) can be prepared following the current procedure: Dissolve 0.1 g agarose in 10 mL of water heated at 363 K stirring it constantly until obtaining a transparent solution. This solution is then passed through a 0.22 μm porosity membrane filter to remove all dust particles or insoluble fibers of agarose. The get-ready gel-solution can be stored into 1.0 mL aliquots in Eppendorf tubes in the fridge. Prior to crystallizing proteins in the agarose, the Eppendorf tube of 1.0 mL can be heated at 363 K by using a heating plate. This is done to melt the gel of 1.0% (*w/v*) concentration. The mixture of precipitant and then agarose will allow to reach a proper temperature to be mixed with the protein avoiding any denaturation problems.

3.2. Electric Field Setup

The growth cell as that shown in Figure 1 was based on that previously published [22] with some modifications for the application of AC currents. It consisted of two polished float conductive indium tin oxide electrodes (usually called ITO) made in glass of dimensions 2.5 × 1.5 cm^2, with a resistance ranging from 4 to 8 Ohms (Delta Technologies, Loveland, CO, USA). The two electrodes (ITO glasses) are placed parallel to each other. The cell is prepared using a rectangular frame as shown in Figure 1 made of elastic black rubber material sealed with vacuum grease to avoid leakage. The closing of the growth cell can be done with a gun for silicone bar melting. The conductive parts of the electrodes surfaces were placed inwards, facing each other. Each of the ITO electrodes is displaced 0.5 cm from one another. This was done to provide the appropriate connection area with the electric alligators to the electrodes (anode/cathode), when an AC current was applied in wave pulse mode. Each cell had a volume capacity of approximately 100 μL by using the batch method.

The batch crystallization conditions for the studied protein must be known before applying the AC current experiments. After closing the growth cell with a cover of melted silicone (Figure 1), the system is connected to the AC apparatus that supplies an alternant current in a pulse-wave mode for testing the four types of waves (at 2 Hz each). The applied AC during the nucleation had to be turned off after 48 h to fix the nuclei on the surface of the ITO electrodes. From here on, the AC growth cell had to be kept at a constant temperature for the rest of the experiment, in order to reach the equilibrium in a maximum time of three days. The device used for these alternating current experiments was a KWD-808 I Multi-purpose health device (usually recommended for acupuncture treatment). One of the purposes of using AC currents for the crystallization of proteins in solution was to demonstrate the efficiency of the procedure in obtaining high-quality protein crystals using all types of precipitants (salts, organic solvents, polyethylene glycols, etc.) and the second is to make it reproducible anywhere. There are two possibilities for growing high-quality protein crystals, one is the mentioned AC current in the breathing-wave pulse mode, and the other is the growing of crystals in agarose inside the strong magnetic field applying radio frequency pulses. However, this second method has a limitation when using polyethylene glycols (PEGs) and agarose together (we must remember that a considerable number of proteins crystallize in PEGs); however, polyethylene glycols will not allow the polymerization of agarose properly.

3.3. Magnetic Field Setup

The setup and the sequence of the experimental steps (from obtaining crystals up to the 3D structure) applying the magnetic field are shown in Figure 6. This setup was based on our previous publication with some modifications [16]. In this case, we have additionally introduced a gel-growth with radio frequency pulses at different angles. For these experiments, the batch crystallization conditions are needed in advance. Once sealed, the capillary pipettes were introduced into an NMR glass tube (5 mm in diameter) and left for at least 68 h under the presence of a magnetic field of 16.5 Tesla (Brucker NMR Advance III HD 700 MHz spectrometer equipped with a 5 mm broadband probe head and a variable temperature unit (VTU). The standard pulse sequence for proton NMR acquisition was used to apply pulses of radio frequencies (56,320 cycles) that systematically vary by 90° the phase of pulses (0, 270°, 270°, 0°, 90°, 180°, 180°, 90°). Each cycle was a rectangular radiofrequency with a gated pulse width of 0.43 μs followed by a delay time of 4.3 s. All experiments were performed at the temperature of 293 K controlled by a unit of the VTU of the NMR spectrometer. After finishing the experiment, the NMR tube was recovered from the magnet, and the capillary pipettes were carefully extracted from the NMR tube. All crystals were immediately mounted and flash-cooled for X-ray data collection.

3.4. X-ray Data Diffraction and Data Processing

Once the protein crystals were harvested from the solution or from the gel, they were cryo-protected for X-ray data collection. For lysozyme, a mixture of 30% (*v*/*v*) glycerol or PEG-1000 with mother liquor of NaCl (precipitating agent) can be used as cryo-protectant. The control crystals grown from the solution reached dimensions of $0.25 \times 0.25 \times 0.25$ mm after three days, while crystals grown in gel reached $0.35 \times 0.35 \times 0.35$ mm in about the same time. X-ray diffraction datasets were collected using the in-house Rigaku/MSC Micromax-007 HF diffractometer, with a rotating anode generator and a DECTRIS-PILATUS 3R/200K-A detector, under cryogenic conditions at 100 K. The crystal-to-detector distance was 40 mm, and 2θ was set to get maximum resolution. Data collection strategies included high redundancy data and each sample was rotated about its omega axis in 0.25° increments. The HKL3000 suite [47] was used to process, merge and scale all datasets.

Initial phases for all the tetragonal (P4$_3$2$_1$2) lysozyme crystals were determined by the molecular replacement method with the program MOLREP [48] (in HKL-3000), using the atomic coordinates of hen egg-white lysozyme determined at a resolution of 1.45 Å (PDB 5T3F), after ligands, alternative conformations, hetero atoms and water molecules were removed. The initial molecular replacement was done at 3.0 Å and rigid-body refinement was performed at the 1.7 Å [49]. Map inspection and model building were done with Coot [50] and the resulting models were refined with Refmac (CCP4) [49] against the full dataset up to the maximum resolution for each crystal. The asymmetric unit of the tetragonal space group P4$_3$2$_1$2 was composed of one chain. The stereochemistry of the structures was checked with PROCHECK [51] and the Ramachandran plot. The processing, scaling and final structures statistics are given in Table 1. Figure 6A–D were produced using the program PyMOL [52].

4. Conclusions

Though crystals grown in gels are usually of a very high quality, unfortunately the agarose cannot be polymerized in the presence of polyethylenglycols (PEGs), which is a major issue as PEGs are a popular precipitating agent for obtaining high-quality crystals. An alternative possibility is the use of magnetic fields with pulses inside to induce the nucleation. Although these types of experiments could give us very good results, they are very expensive to perform. The experiment using AC in solution in a pulse-wave mode is a practical way for obtaining high-quality protein crystals in solution using a variety of precipitating agents without any limitations. Our next research will be focused on the application of this procedure to different types of proteins with different space groups at different crystallization conditions.

Based on these results, we can finally conclude, that the fourth type of pulse wave called "breathing wave", was the most promising one for producing high-quality single crystals in solution (as shown in the X-ray crystallographic analysis). However, further experiments applying this procedure to a variety of proteins from soluble to membrane proteins, need to be performed.

Acknowledgments: One of the authors (Abel Moreno) acknowledges the support of DGAPA UNAM project PAPIIT No. IT200215 for this research. Adela Rodriguez-Romero acknowledges the support of CONACYT (Grant 221169). Carina Pareja-Rivera acknowledges the sponsorship by Sistema Nacional de Investigadores (SNI) CONACYT by the scholarship as an assistant of researcher level 3. All authors acknowledge LANEM-IQ UNAM for the X-ray facilities and data collection help to Ms. Georgina Espinosa. This study made use of UNAM's NMR Lab: LURMN at IQ-UNAM, which is funded by CONACYT Mexico (Project: 0224747). Abel Moreno acknowledges Antonia Sanchez for the English revision of this manuscript.

Author Contributions: Adela Rodriguez-Romero analyzed the crystallographic data; Carina Pareja-Rivera performed the experiments and the analyses of crystal size; Nuria Esturau-Escofet contributed with NMR facilities and the idea of applying pulses; Adela Rodriguez-Romero and Abel Moreno conceived and designed the experiments as well as wrote the manuscript.

Conflicts of Interest: The authors declare no conflict of interest. Additionally, we state that "The founding sponsors had no role in the design of the study; in the collection, analyses, or interpretation of data; in the writing of the manuscript, and in the decision to publish the results".

References

1. Niimura, N.; Bau, R. Neutron protein crystallography: Beyond the folding structure of biological macromolecules. *Acta Crystallogr. A* **2008**, *64*, 12–22. [CrossRef] [PubMed]

2. Blakeley, M.P.; Langan, P.; Niimura, N.; Podjarny, A. Neutron crystallography: Opportunities, challenges, and limitations. *Curr. Opin. Struct. Biol.* **2008**, *18*, 593–600. [CrossRef] [PubMed]

3. Myles, D.A.A.; Dauvergne, F.; Blakeley, M.P.; Meilleur, F. Neutron protein crystallography at ultra-low (<15 k) temperatures. *J. Appl. Cryst.* **2012**, *45*, 686–692. [CrossRef]

4. Yokoyama, T.; Ostermann, A.; Mizuguchi, M.; Niimura, N.; Schader, T.E.; Tanaka, I. Crystallization and preliminary neutron diffraction experiment of human farnesyl pyrophosphate synthase complexed with risedronate. *Acta Crystallogr. F Struct. Biol. Commun.* **2014**, *70*, 470–472. [CrossRef] [PubMed]

5.   Blakeley, M.P.; Hasnain, S.S.; Antonyuk, S.V. Sub-atomic resolution X-ray crystallography and neutron crystallography: Promise, challenges and potential. *IUCrJ* **2015**, *2*, 464–474. [CrossRef] [PubMed]

6.   Baxter, E.L.; Aguila, L.; Alonso-Mori, R.; Barnes, C.O.; Bonagura, C.A.; Brehmer, W.; Brunger, A.T.; Calero, G.; Caradoc-Davis, T.T.; Chaterjee, R.; et al. High-density grids for efficient data collection from multiple crystals. *Acta Crystallogr. D Struct. Biol.* **2016**, *72*, 2–11. [CrossRef] [PubMed]

7.   Cohen, A.E.; Soltis, S.M.; González, A.; Aguila, L.; Alonso-Mori, R.; Barnes, C.O.; Baxter, E.L.; Brehmer, W.; Brewster, A.S.; Brunger, A.T.; et al. Goniometer-based femtosecond crystallography with X-ray free electron lasers. *Proc. Natl. Acad. Sci. USA* **2014**, *111*, 17122–17127. [CrossRef] [PubMed]

8.   Calero, G.; Cohen, A.E.; Luft, J.R.; Newman, J.; Snell, E.H. Identifying, studying and making good use of macromolecular crystals. *Acta Crystallogr. F Struct. Biol. Commun.* **2014**, *70*, 993–1008. [CrossRef] [PubMed]

9.   Boutet, S.; Lomb, L.; Williams, G.J.; Barends, T.R.M.; Aquil, A.; Doak, R.B.; Weierstall, U.; DePonte, D.P.; Steinbrener, J.; Shoeman, R.L.; et al. High-Resolution Protein Structure Determination by Serial Femtosecond Crystallography. *Science* **2012**, *337*, 362–364. [CrossRef] [PubMed]

10. Hunter, M.S.; Fromme, P. Toward structure determination using membrane-protein nanocrystals and microcrystals. *Methods* **2011**, *55*, 387–404. [CrossRef] [PubMed]

11. Fromme, R.; Ishchenko, A.; Metz, M.; Chowdhury, S.R.; Basu, S.; Boutet, S.; Fromme, P.; White, T.A.; Barty, A.; Spence, J.C.H. Serial femtosecond crystallography of soluble proteins in lipidic cubic phase. *IUCrJ* **2015**, *2*, 545–551. [CrossRef] [PubMed]

12. Takayama, Y.; Inui, Y.; Sekiguchi, Y.; Kobayashi, A.; Oroguchi, T.; Yamamoto, M.; Matsunaga, S.; Nakasako, M. Coherent X-ray Diffraction Imaging of Chloroplasts from Cyanidioschyzon merolae by Using X-ray Free Electron Laser. *Plant Cell Physiol.* **2015**, *56*, 1272–1286. [CrossRef] [PubMed]

13. Sazaki, G.; Yoshida, E.; Komatsu, H.; Nakada, T.; Miyashita, S.; Watanabe, K. Effects of a magnetic field on the nucleation and growth of protein crystals. *J. Cryst. Growth* **1997**, *173*, 231–234. [CrossRef]

14. Wakayama, N.I. Effects of a Strong Magnetic Field on Protein Crystal Growth. *Cryst. Growth Des.* **2003**, *3*, 17–24. [CrossRef]

15. Yin, D.-C. Protein crystallization in a magnetic field. *Prog. Cryst. Growth Charact. Mater.* **2015**, *61*, 1–26. [CrossRef]

16. Pareja-Rivera, C.; Cuéllar-Cruz, M.; Esturau-Escofet, N.; Demitri, N.; Polentarutti, M.; Stojanoff, V.; Moreno, A. Recent advances in the Understanding of the Influence of Electric and Magnetic Fields on Protein Crystal Growth. *Cryst. Growth Des.* **2017**, *17*, 135–145. [CrossRef]

17. Yan, E.K.; Zhang, C.Y.; He, J.; Yin, D.-C. An Overview of Hardware for Protein Crystallization in a Magnetic Field. *Int. J. Mol. Sci.* **2016**, *17*, E1906. [CrossRef] [PubMed]

18. Gavira, J.A.; García-Ruiz, J.M. Effects of a Magnetic Field on Lysozyme Crystal Nucleation and Growth in a Diffusive Environment. *Cryst. Growth Des.* **2009**, *9*, 2610–2615. [CrossRef]

19. Hammadi, Z.; Veesler, S. New approaches on crystallization under electric fields. *Prog. Biophys. Mol. Biol.* **2009**, *101*, 38–44. [CrossRef] [PubMed]

20. Koizumi, H.; Uda, S.; Fujiwara, K.; Tachibana, M.; Kojima, K.; Nozawa, J. Crystallization of high-quality protein crystals using an external electric field. *J. Appl. Cryst.* **2015**, *48*, 1507–1513. [CrossRef]

21. Sazaki, G.; Moreno, A.; Nakajima, K. Novel coupling effects of the magnetic and electric fields on protein crystallization. *J. Cryst. Growth* **2004**, *262*, 499–502. [CrossRef]

22. Gil-Alvaradejo, G.; Ruiz-Arellano, R.R.; Owen, C.; Rodríguez-Romero, A.; Rudiño-Piñera, E.; Antwi, M.K.; Stojanoff, V.; Moreno, A. Novel Protein Crystal Growth Electrochemical Cell For Applications In X-ray Diffraction and Atomic Force Microscopy. *Cryst. Growth Des.* **2011**, *11*, 3917–3922. [CrossRef]

23. Nieto-Mendoza, E.; Frontana-Uribe, B.A.; Sazaki, G.; Moreno, A. Investigations on electromigration phenomena for protein crystallization using crystal growth cells with multiple electrodes: Effect of the potential control. *J. Cryst. Growth* **2005**, *275*, e1437–e1446. [CrossRef]

24. Frontana-Uribe, B.A.; Moreno, A. On Electrochemically Assisted Protein Crystallization and Related Methods. *Cryst. Growth Des.* **2008**, *8*, 4194–4199. [CrossRef]

25. Pérez, Y.; Eid, D.; Acosta, F.; Marin-Garcia, L.; Jakoncic, J.; Stojanoff, V.; Frontana-Uribe, B.A.; Moreno, A. Electrochemically Assisted Protein Crystallization of Commercial Cytochrome *c* without Previous Purification. *Cryst. Growth Des.* **2008**, *8*, 2493–2496. [CrossRef]

26. Koizumi, H.; Uda, S.; Fujiwara, K.; Nozawa, J. Control of Effect on the Nucleation Rate for Hen Egg White Lysozyme Crystals under Application of an External AC Electric Field. *Langmuir* **2011**, *27*, 8333–8338. [CrossRef] [PubMed]

27. Flores-Hernandez, E.; Stojanoff, V.; Arreguin-Espinosa, R.; Moreno, A.; Sanchez-Puig, N. An electrically assisted device for protein crystallization in a vapor-diffusion setup. *J. Appl. Crystallogr.* **2013**, *46*, 832–834. [CrossRef] [PubMed]

28. Koizumi, H.; Uda, S.; Fujiwara, K.; Tachibana, M.; Kojima, K.; Nozawa, J. Improvement of crystal quality for tetragonal hen egg white lysozyme crystals under application of an external alternating current electric field. *J. Appl. Crystallogr.* **2013**, *46*, 25–29. [CrossRef]

29. Uda, S.; Koizumi, H.; Nozawa, J.; Fujiwara, K. Crystal Growth under External Electric Fields. In Proceedings of the 2014 International Conference of Computational Methods in Sciences and Engineering (ICCMSE 2014), Athens, Greece, 4–7 April 2014; Volume 1618, pp. 261–264.

30. De la Mora, E.; Flores-Hernandez, E.; Jakoncic, J.; Stojanoff, V.; Siliqi, D.; Sanchez-Puig, N.; Moreno, A. SdsA polymorph isolation and improvement of their crystal quality using nonconventional crystallization techniques. *J. Appl. Crystallogr.* **2015**, *48*, 1551–1559. [CrossRef]

31. Nanev, C.N.; Penkova, A. Nucleation of lysozyme crystals under external electric and ultrasonic fields. *J. Cryst. Growth* **2001**, *232*, 285–293. [CrossRef]

32. Nanev, C.N.; Penkova, A. Nucleation and growth of lysozyme crystals under external electric field. *Coll. Surf. A Physicochem. Eng. Asp.* **2002**, *209*, 139–145. [CrossRef]

33. Penkova, A.; Gliko, O.; Dimitrov, I.L.; Hodjaoglu, F.V.; Nanev, C.; Vekilov, P.G. Enhancement and suppression of protein crystal nucleation due to electrically driven convection. *J. Cryst. Growth* **2005**, *275*, e1527–e1532. [CrossRef]

34. Taleb, M.; Didierjean, C.; Jelsch, C.; Mangeot, J.P.; Capelle, B.; Aubry, A. Crystallization of proteins under an external electric field. *J. Cryst. Growth* **1999**, *200*, 575–582. [CrossRef]

35. Taleb, M.; Didierjean, C.; Jelsch, C.; Mangeot, J.P.; Aubry, A. Equilibrium kinetics of lysozyme crystallization under an external electric field. *J. Cryst. Growth* **2001**, *232*, 250–255. [CrossRef]

36. Yoshikawa, H.Y.; Murai, R.; Sugiyama, S.; Sazaki, G.; Kitatani, T.; Takahashi, Y.; Adachi, H.; Matsumura, H.; Murakami, S.; Inoue, T.; et al. Femtosecond laser-induced nucleation of protein in agarose gel. *J. Cryst. Growth* **2009**, *311*, 956–959. [CrossRef]

37. Yoshikawa, H.Y.; Murai, R.; Adachi, H.; Sugiyama, S.; Maruyama, M.; Takahashi, Y.; Takano, K.; Matsumura, H.; Inoue, T.; Murakami, S.; et al. Laser ablation for protein crystal nucleation and seeding. *Chem. Soc. Rev.* **2014**, *43*, 2147–2158. [CrossRef] [PubMed]

38. Gavira, J.A.; Garcia-Ruiz, J.M. Agarose as crystallisation media for proteins II: Trapping of gel fibres into the crystals. *Acta Crystallogr. D Biol. Crystallogr.* **2002**, *D58*, 1653–1656. [CrossRef]

39. Charron, C.; Robert, M.C.; Capelle, B.; Kadri, A.; Jenner, G.; Giegé, R.; Lorber, B. X-ray diffraction properties of protein crystals prepared in agarose gel under hydrostatic pressure. *J. Cryst. Growth* **2002**, *245*, 321–333. [CrossRef]

40. Sauter, C.; Balg, C.; Moreno, A.; Dhouib, K.; Theobald-Dietrich, A.; Chenevert, R.; Giege, R.; Lorber, B. Agarose gel facilitates enzyme crystal soaking with a ligand analog. *J. Appl. Crystallogr.* **2009**, *42*, 279–283. [CrossRef]

41. Gonzalez-Ramirez, L.A.; Caballero, A.G.; Garcia-Ruiz, J.M. Investigation of the Compatibility of Gels with Precipitating Agents and Detergents in Protein Crystallization Experiments. *Cryst. Growth Des.* **2008**, *8*, 4291–4296. [CrossRef]

42. Gavira, J.A.; Van Driessche, A.E.S.; Garcia-Ruiz, J.M. Growth of Ultrastable Protein-Silica Composite Crystals. *Cryst. Growth Des.* **2013**, *13*, 2522–2529. [CrossRef]

43. Choquesillo-Lazarte, D.; Garcia-Ruiz, J.M. Poly (ethylene) oxide for small-molecule crystal growth in gelled organic solvents. *J. Appl. Crystallogr.* **2011**, *44*, 172–176. [CrossRef]

44. Pietras, Z.; Lin, H.-T.; Surade, S.; Luisi, B.; Slattery, O.; Pos, K.M.; Moreno, A. The use of novel organic gels and hydrogels in protein crystallization. *J. Appl. Crystallogr.* **2010**, *43*, 58–63. [CrossRef]

45. Chayen, N.E.; Saridakis, E. Protein crystallization: From purified protein to diffraction-quality crystal. *Nat. Methods* **2008**, *5*, 147–153. [CrossRef] [PubMed]

46. Saijo, Sh.; Yamada, Y.; Sato, T.; Tanaka, N.; Matsui, T.; Sazaki, G.; N.akajima, K.; Matsuura, Y. Structural consequences of the hen egg-white lysozyme orthorhombic crystal growth in a high magnetic field: Validation of X-ray diffraction intensity, conformational energy searching and quantitative analysis of B factors and mosaicity. *Acta Crystallogr. D Biol. Crystallogr.* **2005**, *D61*, 207–217. [CrossRef] [PubMed]

47. Minor, W.; Cymborowski, M.; Otwinowski, Z.; Chruszcz, M. HKL-3000: The integration of data reduction and structure solution-from diffraction images to an initial model in minutes. *Acta Crystallogr. D Biol. Crystallogr.* **2006**, *D62*, 859–866. [CrossRef] [PubMed]

48. Vagin, A.; Teplyakov, A. Molecular replacement with MOLREP. *Acta Crystallogr.* **2010**, *D66*, 22–25.

49. Mushudov, G.N.; Vagin, A.A.; Dodson, E.J. Refinement of macromolecular structures by the maximum-likelihood method. *Acta Crystallogr. D Biol. Crystallogr.* **1997**, *D53*, 240–255. [CrossRef] [PubMed]

50. Emsley, P.; Cowtan, K. Coot: Model-building tools for molecular graphics. *Acta Crystallogr. D Biol. Crystallogr.* **2004**, *60*, 2126–2132. [CrossRef] [PubMed]

51. Laskowski, R.A.; MacArthur, M.W.; Thornton, J.M. PROCHECK: Validation of protein structure coordinates. In *International Tables of Crystallography, Volume F. Crystallography of Biological Macromolecules*; Rossmann, M.G., Arnold, E., Eds.; Kluwer Academic Publishers: Dordrecht, The Netherlands, 2001; pp. 722–725.

52. De Lano, W.L. *The PyMOL Molecular Graphics System*; DeLano Scientific LLC: San Carlos, CA, USA, 2002.

crystals

MDPI

Article

A Graphene-Based Microfluidic Platform for Electrocrystallization and In Situ X-ray Diffraction

Shuo Sui, Yuxi Wang, Christos Dimitrakopoulos and Sarah L. Perry *

Department of Chemical Engineering, University of Massachusetts Amherst, MA 01003, USA;
ssui@umass.edu (S.S.); yuxiwang@engin.umass.edu (Y.W.); dimitrak@umass.edu (C.D.)
* Correspondence: perrys@engin.umass.edu

Received: 16 December 2017; Accepted: 30 January 2018; Published: 1 February 2018

Abstract: Here, we describe a novel microfluidic platform for use in electrocrystallization experiments. The device incorporates ultra-thin graphene-based films as electrodes and as X-ray transparent windows to enable in situ X-ray diffraction analysis. Furthermore, large-area graphene films serve as a gas barrier, creating a stable sample environment over time. We characterize different methods for fabricating graphene electrodes, and validate the electrical capabilities of our device through the use of methyl viologen, a redox-sensitive dye. Proof-of-concept electrocrystallization experiments using an internal electric field at constant potential were performed using hen egg-white lysozyme (HEWL) as a model system. We observed faster nucleation and crystal growth, as well as a higher signal-to-noise for diffraction data obtained from crystals prepared in the presence of an applied electric field. Although this work is focused on the electrocrystallization of proteins for structural biology, we anticipate that this technology should also find utility in a broad range of both X-ray technologies and other applications of microfluidic technology.

Keywords: microfluidics; electrocrystallization; protein crystallization; in situ diffraction; serial crystallography

1. Introduction

The application of both internal and external electric fields has long been shown to modulate the rate of protein crystallogenesis, and serves as a possible tool for enhancing crystal quality [1–5]. The presence of an electric field during protein crystallization has been shown to affect both the rate of nucleation and the rate of crystal growth by controlling the local concentration and concentration gradients of proteins and the associated crystallization reagents [1–14]. These methods have the potential to improve the success rate associated with protein crystallogenesis and enhance our understanding of the structure–function relationship in challenging biomacromolecular targets.

Electrocrystallization platforms have been reported for a variety of crystallization schemes, including batch [6–11,14–20], vapor diffusion [5,13–16], and counter-diffusion [21], taking advantage of both internal [6,7,12–14,19–22] and external electric fields [5,8–10,15,18,23]. These strategies have also explored the effects of constant (DC) [5–7,13–17,20–22] and alternating (AC) electric fields [8–12,16–18]. However, the benefits observed for the various electrode arrangements and crystallization setups reported to date have been limited by the need to manually harvest crystals for subsequent diffraction analysis.

Microfluidic and microscale devices have a demonstrated potential to enable both protein crystallization and in situ X-ray diffraction. Such platforms have been increasingly harnessed to facilitate the diffraction analysis of challenging targets for both static and dynamic structure determination. Various platforms have been developed to improve the growth and subsequent mounting of tiny and fragile crystals for X-ray diffraction analysis [24–28], including dense array-style devices [29–41], platforms for the lipidic cubic phase crystallization of membrane proteins [42,43], and thin-film sandwich devices [44]. In the meantime, the challenges of such platforms lie in the need to maintain a protected sample environment, as well as minimize the interference of device materials with the subsequent X-ray analysis. To address these two issues, we recently developed a microfluidic device architecture that takes advantage of large-area sheets of graphene [45]. The use of atomically-thin graphene films minimizes the amount of material surrounding a crystal, while serving as a vapor-diffusion barrier that is stable against significant water loss over the course of weeks. This approach enables the incubation of protein crystallization trials and direct in situ analysis of the resulting crystals. Here, we harness the intrinsic conductivity of graphene [46,47] to enable electrocrystallization experiments in the precisely controlled microfluidic geometry of our device, along with an in situ X-ray analysis of the resulting crystals.

2. Materials and Methods

2.1. Graphene Film Preparation

Large-area graphene was synthesized on a copper substrate (Graphene Platform, Tokyo, Japan) by chemical vapor deposition in a quartz tube furnace (Planar Tech, The Woodlands, TX, USA) using standard methods [48–51]. After synthesis, the back side of the copper substrate was scrubbed with a Kimwipe to remove residual graphene. Patterning of the graphene electrodes was achieved using two different methods. The first method simply used thin-tip tweezers (TDI International Inc., Tucson, AZ, USA) to scratch a narrow line into the graphene/copper film. The second method defined the desired structure of the electrodes using a protective mask made from a piece of thermal release tape (Semiconductor Equipment Corp., Moorpark, CA, USA) cut to the desired shape using a cutting plotter (Graphtech CE6000, Irvine, CA, USA), followed by a five-minute etching of the exposed graphene by an oxygen plasma (Harrick Plasma, Ithaca, NY, USA). Following the patterning of the graphene electrodes, a roughly 500-nm thick layer of poly(methylmethacrylate) (950PMMA A4, Microchem, Westborough, MA, USA) was then spin coated (Specialty Coating Systems, Amherst, NH, USA) onto the graphene at 1000 rpm to serve as a support layer. The PMMA film was cured at 120 °C for 10 min. The resulting PMMA/graphene film was then released from the copper substrate by back-etching of the copper in an aqueous solution of $FeCl_3$ copper etchant (Transene, Danvers, MA, USA) for 3 h, followed by three rinse cycles in MilliQ water (18.2 MΩ-cm, Millipore Inc., Billerica, MA, USA). The graphene film floats on the surface of water, and was transferred directly onto the target substrate by lifting it from the water surface. The assembled layers were then allowed to dry at room temperature.

2.2. Device Architecture

The structure of the microfluidic platform was designed to enable the application of an internal electrical field to the crystallization solution through patterned graphene electrodes (Figure 1). The overall device structure was assembled around a chamber cut into a 100-μm double-sided adhesive-backed polyester film (Adhesive Research #90668, Glen Rock, PA, USA) using a cutting plotter (Graphtec CE6000, Irvine, CA, USA). The layer containing the patterned graphene/PMMA electrode was adhered onto this film with the gap in the electrodes located near the center of the chamber. A supporting frame of cyclic olefin copolymer (COC, Topas Advanced Polymers, Florence, KY, USA) with window structures aligned to the crystallization chamber was adhered to the outside of the notched graphene/PMMA film to provide mechanical stability. After filling of the device, the chamber was sealed with a top layer containing a smaller, unmodified graphene/PMMA film, supported on a COC frame. In contrast to the electrode layer, the top graphene/PMMA film was oriented with the PMMA layer facing the crystallization chamber so that the graphene would not contribute to the conductivity of the cell. Finally, the small side features cut into the polyester film and the top support layer of COC were filled by a gallium–indium liquid alloy (Sigma Aldrich, St. Louis, MO, USA) to create the electrode contact between graphene films and the electrode needles running to a power supply (Figure 2). It should be noted that the thickness of the adhesive layer used to define the crystallization chamber can be modulated to match the size of the resulting crystals, and minimize the amount of excess liquid surrounding the crystals during data collection.

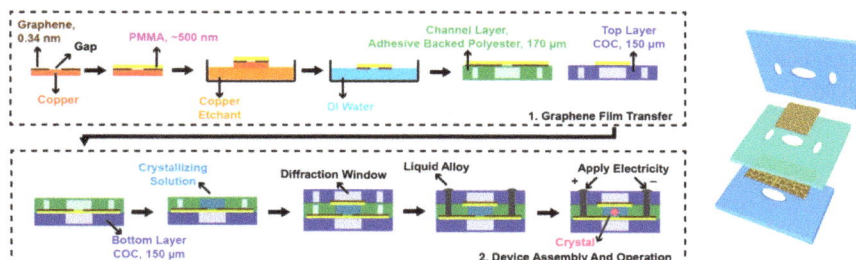

Figure 1. Schematic illustration of the fabrication scheme and device architecture for thin-film graphene-based microfluidics. (**1**) A patterned graphene film on copper is first coated with a layer of poly(methylmethacrylate) (PMMA), and then released from the copper substrate by etching. The subsequent film is floated on the surface of water for transfer to either an adhesive-backed polyester film that defines the crystallization chamber, or a cyclic olefin copolymer film to form the top layer of the device. (**2**) Assembly of the device proceeded with the application of a COC bottom layer to the crystallization layer assembly to provide additional stability. Following the addition of the crystallization solution, the device is then sealed with the COC top layer. Electrical contact to the graphene electrodes is made using a liquid alloy, and the electrocrystallization experiment can take place.

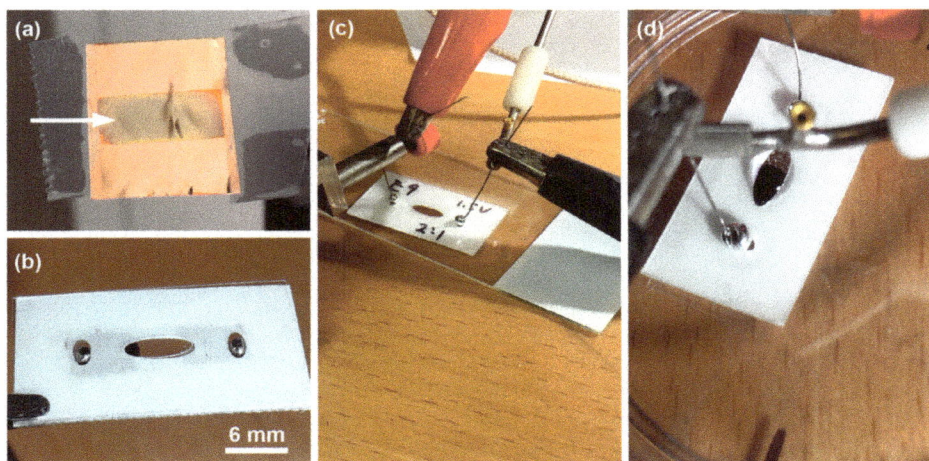

Figure 2. (**a**) Graphene film on copper growth substrate after oxygen plasma treating. The brassy yellow area in the middle of the film was exposed to the plasma, while the upper and lower regions were covered and protected by a mask created from thermal release tape. (**b**) A view of the patterned electrodes (regions of light grey) in an assembled device. (**c**) System setup for electrocrystallization experiments. Alligator clips attached to metal needles and gallium–indium alloy droplets were used to make electrical contact with the integrated graphene electrodes. (**d**) The presence of an applied electric field can be observed visually based on the color change of methyl viologen from clear to purple as it undergoes reduction at the cathode.

2.3. Electrode Characterization

To quantitatively characterize the film electrical resistance at different conditions, we measured the voltage resulting from a current sweep from 0 to 100 µA using a semiconductor characterization system (Keithley 4200 SCS, Tektronix Inc., Beaverton, OR, USA) on intact graphene films and patterned graphene electrodes in air, and in the presence of a crystallization solution (Figure 3). We compared electrodes patterned by both physical abrasion and plasma etching. The corresponding electrical resistances were calculated based on the resultant voltage–current relationships and the device architecture. All of the tests were performed in triplicate.

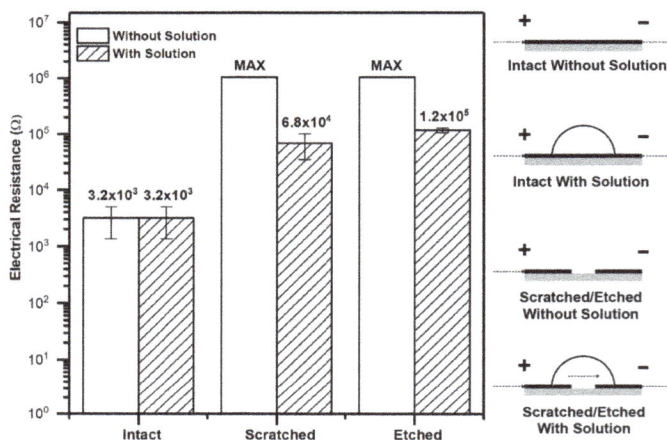

Figure 3. Plot of the measured average electrical resistance of an intact graphene film, a graphene film where electrodes were fabricated by physical scratching, and a graphene film where the electrode structure was created by plasma etching. Data are shown for both the electrode structure alone (without solution, clear bars) and for a device filled with crystallization solution (hatched bars), and are the average of measurements from three separate devices. The maximum resistance measured for the two electrode structures in air suggests an infinite resistance, beyond the range of the instrument.

2.4. Redox Chemistry Testing

Methyl viologen (MV) is a redox and oxygen-sensitive dye. A solution of methyl viologen (Sigma Aldrich, St. Louis, MO, USA) in water was prepared at 150 mM. In the presence of oxygen, methyl viologen is present as fully oxidized MV^{2+}, resulting in a colorless solution. The partially oxidized MV^{+} species is a brilliant purple, while the fully reduced, neutral MV^{0} is typically light yellow [52–54]. To test the ability of our graphene-based devices to conduct electricity and drive redox chemistry, 10 µL of fully oxidized methyl viologen was placed onto a patterned graphene electrode. An applied voltage was then slowly increased from 0 V to 3 V, and then held at 3 V for 5 min, during which time the subsequent color changes were observed (Figure 4).

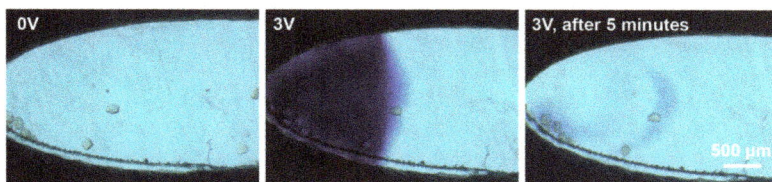

Figure 4. Optical micrographs of an electrocrystallization device containing 150 mM of methyl viologen (MV) and crystals of lysozyme under the influence of 0 V, 3 V, and 3 V at longer times. The initial color change from clear to purple is the result of the reduction of MV^{2+} to MV^+, while the subsequent loss of color at higher voltages and longer times is due to the further reduction from MV^+ to MV^0. The pale yellow color of the MV^0 was difficult to discern compared to the fully oxidized MV^{2+} species, because of the small path length in our microfluidic devices. Color changes were only observed in the vicinity of the cathode.

2.5. Protein Crystallization and X-ray Diffraction

Hen egg white lysozyme (HEWL, Hampton Research Inc., Aliso Viejo, CA, USA) was prepared in 50 mM of sodium acetate (Fisher) and 20% (*w/v*) glycerol (Fisher Scientific, Hampton, NH, USA) with a concentration of 80 mg/mL at pH 4.8. The protein solution was then fully mixed by vortexing with a precipitant solution containing 0.68 M of sodium chloride (Sigma, St. Louis, MO, USA) and 50 mM of sodium acetate at pH 4.8 at a volumetric ratio of 2:3. All of the solutions were filtered before use through a 0.2-μm membrane (Millipore, Billerica, MA, USA) to remove impurities. Crystallization was performed using a microbatch-type method [55]; 3.2 μL of mixed solution was pipetted immediately after preparation, and sealed into the device. It should be noted that the volume of solution added to the device should be carefully controlled to match the volume of the chamber, as excess liquid will be squeezed out of the chamber, and will adversely affect device sealing.

The crystallization experiment was performed in a 4 °C cold room under different applied voltages using a potentiostat (Arksen 305-2D, City of Industry, CA, USA). For a given experiment, simultaneous tests were performed on multiple devices at the applied voltage, alongside a control device with no applied voltage. Crystal growth was monitored hourly using a stereomicroscope (Zeiss SteREO Discovery V12, Oberkochen, Germany) under cross-polarized light (Figure 5a and Figure S1). After crystallization was complete, the devices were disconnected from the voltage supply, sealed in Petri dishes (Fisher Scientific, Hampton, NH, USA), and stored at 4 °C prior to X-ray analysis. Replicate crystallization experiments were performed over a range of applied voltages (0 V to 1.8 V), demonstrating the reproducibility of our approach.

The quantification of crystal size as a function of time was done using the size measurement function in ImageJ software (NIH, Bethesda, MD, USA) [56] by manually outlining crystal edges. Crystals appeared to be randomly oriented. However, the overall aspect ratio of the crystals was similar, allowing for the reasonable use of a calculated two-dimensional area to represent the three-dimensional size of a crystal. For each time point, all of the crystals in each chip were measured, and the average projected area was calculated (Figure 5b,c and Figure S2). Error bars represented the standard deviation from the mean. A comparison between crystal sizes at different time points and voltage conditions was performed using ANOVA. While the data in Figure 5 represent the results of only three individual devices, replicate experiments show similar trends, relative to controls (see Figures S1 and S2).

Immediately after crystallization, the chips were stored in 4 °C, and analyzed within a couple of days. The chip was mounted on the goniometer using a custom magnetic mounting base (Crystal Positioning Systems, Jamestown, NY, USA). The X-ray system (Rigaku XtalAB PRO MM007, Tokyo, Japan) operated at an X-ray wavelength of 1.542 Å and a beam size of ~200 μm, along with a PILATUS3 R 200K detector (Dectris AG, Baden-Dättwil, Switzerland). The chip was initially mounted perpendicular to the beam path. Crystal targeting and focusing and were done by adjusting the goniometer positions. The sample-to-detector distance was set at 40 mm, giving a maximum resolution of 1.95 Å. A 10-s exposure and 1° oscillation were used. Before collecting a complete dataset, the sample orientations corresponding to the first and last frames were tested to avoid overlapping signals from nearby crystals. The collected diffraction patterns were then analyzed using the HKL 3000 software package (HKL Research Inc., Charlottesville, VA, USA) for indexing, refinement, integration, and scaling. The X-ray diffraction data extended to the maximum resolution limit of the X-ray setup, showing a signal-to-noise level in the highest resolution shell of $I/\sigma(I) > 3.0$ for all of the samples (Figure 6, Table 1).

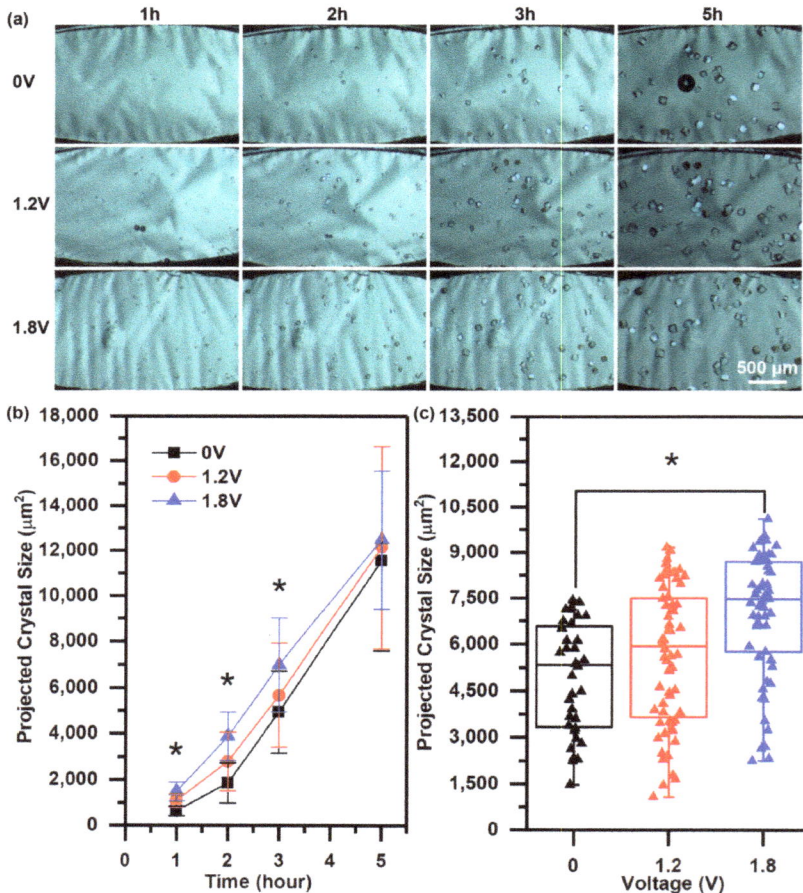

Figure 5. (**a**) Optical micrographs under cross-polarized light showing the time evolution of hen egg-white lysozyme (HEWL) crystal nucleation and growth with the application of 0 V, 1.2 V, and 1.8 V in a microfluidic device. (**b**) A plot of the average crystal size as a function of time from the images in (**a**). Error bars represent the standard deviation. (**c**) A box and whiskers plot of the crystal size distribution at 3 h with the application of voltages at 0 V, 1.2 V, and 1.8 V. The middle line shows the median, and the ends of the box indicate the upper and lower quartiles. * Crystals prepared at 1.8 V at 1 h, 2 h, and 3 h were statistically larger than those prepared at 0 V, ANOVA $p < 0.01$.

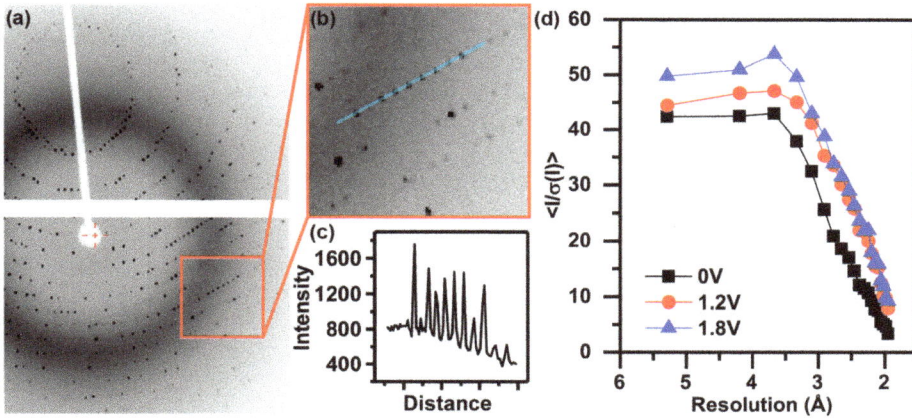

Figure 6. (**a**) A typical obtained X-ray diffraction pattern and (**b**) a magnified view showing details of diffracted spots. (**c**) Pixel intensity along the blue line in the inset indicated the high levels of signal-to-noise observed in the data. (**d**) A plot of signal-to-noise ratios at different resolution shells of diffraction patterns from crystals grown under different voltages.

Table 1. Crystallographic statistics for data obtained using graphene-based microfluidics under different applied voltages.

Parameter	0 V	1.2 V	1.8 V
Data Collection			
Total # Frames	50	55	90
Resolution (Å)	50–1.95 (1.98–1.95)	50–1.95 (1.98–1.95)	50–1.95 (1.98–1.95)
Space Group	$P4_32_12$	$P4_32_12$	$P4_32_12$
Unit Cell (Å)	a = b = 79.35, c = 37.99	a = b = 79.23, c = 38.09	a = b = 78.92, c = 38.19
Single Reflections			
Total Obs.	31,372	34,440	55,551
Unique Obs.	7086	8817	9053
Redundancy	4.4 (3.4)	3.9 (3.2)	6.1 (5.1)
R_{meas} [a]	0.069 (0.399)	0.052 (0.202)	0.076 (0.255)
R_{pim} [b]	0.031 (0.206)	0.025 (0.105)	0.031 (0.112)
$CC_{1/2}$ [c]	0.971 (0.883)	0.990 (0.961)	0.951 (0.890)
$<I/\sigma(I)>$	22.69 (3.39)	33.24 (7.89)	36.98 (9.21)
Completeness (%)	76.0 (82.4)	94.6 (95.4)	97.5 (97.1)

Data in the parenthesis are from the highest resolution shell. [a] $R_{meas} = \frac{\sum_{hkl} \sqrt{\frac{n}{n-1}} \sum_{j=1}^{n} |I_{hkl,j} - <I_{hkl}>|}{\sum_{hkl} \sum_{j} I_{hkl,j}}$, [b] $R_{pim} = \frac{\sum_{hkl} \sqrt{\frac{1}{n-1}} \sum_{j=1}^{n} |I_{hkl,j} - <I_{hkl}>|}{\sum_{hkl} \sum_{j} I_{hkl,j}}$, where I is the reflection intensity and $<I>$ is its average, and $\sqrt{\frac{n}{n-1}}$ and $\sqrt{\frac{1}{n-1}}$ are factors for multiplicity. [c] $CC_{1/2}$ is the Pearson correlation coefficient with the dataset randomly being split in half, and $CC = \frac{\sum (x-<x>)(y-<y>)}{\sqrt{\sum (x-<x>)^2 \sum (y-<y>)^2}}$, where x, y are single samples.

3. Results and Discussion

The goal of this work was to take advantage of atomically-thin, conductive graphene films to enable electrocrystallization experiments in a microfluidic device, followed by in situ X-ray diffraction analysis of the resulting crystals. This work builds on our previously reported graphene-based platform for serial crystallography [45], but requires the fabrication and integration of patterned graphene electrodes, rather than simple graphene windows.

While it is possible to create a set of electrodes by simply adhering two separate pieces of graphene to a substrate, we took advantage of a more controlled fabrication strategy to enable careful control of the electrode spacing and geometry. Here, we used a protective film of thermal release tape to facilitate direct patterning of the graphene, using an oxygen plasma. Following plasma treatment and removal of the protective film, we observed clear patterning of the graphene to reveal the underlying copper substrate (Figure 2a). Thus, the width of the resulting gap can be easily controlled to modulate the electric field strength. Experiments were typically done using a gap size of 3 mm. A roughly 500 nm-thick layer of PMMA was then spin coated onto the graphene/copper surface to facilitate the retention of electrode geometry after removal from the underlying copper substrate and transfer to the target device layers. The graphene electrodes could be observed on the fully assembled devices as areas of light grey color located on the ends of the microfluidic channel, relative to the white background of the middle adhesive channel layer of the device (Figure 2b). This design takes advantage of relatively cheap materials and fabrication strategies, such that the material cost of a single device should be on the order of $1 (USD) or less, depending on economies of scale.

We compared the resulting electrical properties of these plasma-etched graphene electrodes with an analogous electrode layout fabricated by simple physical abrasion (Figure 3). The electrical resistance of an intact graphene film was relatively low, and highly reproducible, as expected for an atomically-thin conductive material. While the effectively infinite resistance measured for the two electrode setups in air clearly demonstrated the separation of the two electrodes, clear differences were observed in the operation of the devices in the presence of crystallization solution. Devices with the electrodes fabricated via physical abrasion showed substantially lower and more variable resistivity values compared with the plasma etching method. The lower resistance observed for the physical abrasion method suggests the presence of graphene residue in the gap area between the electrodes. Thus, while this kind of simplified fabrication scheme can be applied, it has the potential to adversely affect both the performance and reproducibility of the resulting device in electrocrystallization experiments. Subsequent experiments were performed using plasma-etched electrodes.

To further visualize the effectiveness of our devices, we utilized methyl viologen (MV) as a redox-sensitive colorimetric indicator. The solution was observed to change from colorless (MV^{2+}) to brilliant purple (MV^+) near the cathode upon the application of 1 V, consistent with the reported value of the redox potential for the $MV^{2+} + e^- \rightarrow MV^+$ reaction of ~0.7 V (Figure 2) [52–54]. A similar color change was observed for a slurry of lysozyme crystals containing methyl viologen (Figure 4). In both experiments, the observed change in color only occurred in the area of the device defined by the cathode. We hypothesize that the localization of this color change near the cathode is due to an enhancement of the redox reaction by the solid graphene electrode. Increasing the applied voltage to 3 V resulted in an intensification of the observed purple color, due to the increased generation of the MV^+ species. However, after several minutes, the solution transitioned from purple to clear, as MV^+ was further reduced to MV^0 (Figure 4). Again, this result was expected, based on the reported redox potential for the $MV^+ + e^- \rightarrow MV^0$ reaction [52–54]. The pale yellow color of the MV^0 was difficult to discern compared to the fully oxidized MV^{2+} species, because of the small path length in our microfluidic devices. It is also noteworthy that despite the potential for water electrolysis at these applied voltages, the reaction rate on graphene electrodes is relatively slow. Thus, the formation of bubbles was typically not observed during the course of an experiment.

Having demonstrated the electrical performance of our device, we then proceeded to study the electrocrystallization of lysozyme as a function of time. With a 3-mm patterned gap on the graphene film, the applied voltage resulted in an electric field strength in the range of 0.4 V/mm to 0.6 V/mm, which is similar to a range reported in the literature [4]. As shown in Figure 5, Figures S1 and S2, the presence of an applied voltage resulted in an increased rate of protein nucleation and growth, which is consistent with previous literature reports [1–14]. Interestingly, these trends were only significant at short times. For instance, after 1 h, 2 h, and 3 h, the average size of crystals grown under the influence of an applied voltage was statistically different compared to a control sample (Figure 5b,c), while this difference is lost by 5 h.

In addition to the effects on nucleation and growth, we did not observe a significant preferential localization of crystals within the device. This is in contrast to previous reports for the electrocrystallization of lysozyme, where crystals were typically localized near the cathode. We hypothesize that the broad spatial distribution of crystals, as well as the similar crystal size at long times, is a consequence of the relatively short time period over which these experiments were performed. The increased rate of crystal nucleation and growth associated with electrocrystallization is typically associated with the electromigration of the protein, and subsequent increases in concentration near the relevant electrode. Thus, it is possible that the crystallization conditions used here fall very close to the nucleation region such that only minimal increases in the local protein concentration were necessary to facilitate nucleation, while allowing for the appearance of crystal growth throughout the device and the similarity of crystal size at long times.

After crystallization, the devices were stored at 4 °C for several days prior to X-ray diffraction analysis. We collected a room temperature dataset from a representative crystal grown under each of the applied voltage conditions (0 V, 1.2 V, 1.8 V). Data were collected and analyzed to the maximum resolution of our diffraction setup. At this limit of 1.95 Å, the $I/\sigma(I)$, or signal-to-noise level in the highest resolution shell, was above 3.0 for all of the samples, and was significantly higher for those samples prepared in the presence of an electric field, than those without (Figure 6, Table 1). This high signal-to-noise was expected, due to the minimal contributions of the device materials to the level of background noise [45]. The size of the X-ray beam and the presence of nearby crystals limited the number of frames that could be collected from a given sample. While it was possible to collect nearly complete datasets for the 1.2 V and 1.8 V samples, a lower completeness was obtained for the 0 V sample. Despite these differences, the data suggest that the crystals grown in the presence of an electric field may diffract to higher resolution than crystals grown without. Unfortunately, we were unable to confirm this result directly because of the limitations of our X-ray diffraction setup. It should be noted that, although the crystal size varied between the various voltage conditions early on, this difference was lost over longer time periods. Care was taken to select crystals of similar size. Thus, the difference in the observed signal-to-noise should not be a consequence of differences in crystal size. These results agree with previous literature reports where a higher signal-to-noise [7] was observed for crystals grown in an electric field. It is important to note that this is the first report where direct, in situ measurements of the X-ray diffraction quality could be obtained on protein crystals grown via electrocrystallization, without the need for handling of fragile capillaries [16] or the use of hard X-rays to limit absorption from the crystallization cell [14].

4. Summary

In conclusion, we have presented a straightforward method for the incorporation of graphene-based electrodes into an ultra-thin, X-ray compatible microfluidic platform. We have demonstrated the utility of this setup to enable in situ X-ray diffraction data collection for electrocrystallization experiments. Our data agree with previous reports, showing faster crystal nucleation and an improvement in signal-to-noise for crystals grown in the presence of an electric field. Building on these results, our microfluidic approach has the potential to enable high-throughput analysis of a tremendous range of crystallization and electric field conditions to better map out the effect of these parameters on crystal quality in general. This approach is also amenable to serial crystallography experiments, where our microfluidic array chip could be used to grow hundreds or thousands of microcrystals for serial diffraction analysis. Looking beyond structural biology, the integration of ultra-thin graphene electrodes into microfluidic devices could similarly enable powerful high-throughput experiments in a range of other fields.

Supplementary Materials: The following are available online at www.mdpi.com/2073-4352/8/2/76/s1. Additional electrocrystallization results are available in Figures S1 and S2.

Acknowledgments: We acknowledge support from the NSF Science and Technology Center on Biology with X-ray Lasers (NSF-1231306). X-ray diffraction data were obtained at the University of Massachusetts Amherst Institute for Applied Life Sciences Biophysical Characterization Facility. We would like to acknowledge Lizz Bartlett and Derrick Deming for assistance with X-ray diffraction experiments, Jungwoo Lee and Jun-Goo Kwak for help with and the use of the cutting plotter, Whitney Blocher, Ryan Carpenter and Xiangxi "Zoey" Meng for help with statistical analysis, and Scott Garman and Yiliang Zhou for valuable discussions.

Author Contributions: Shuo Sui, Yuxi Wang and Sarah L. Perry designed and performed the experiments. Shuo Sui and Sarah L. Perry analyzed the data. All authors helped with writing the manuscript.

Conflicts of Interest: The authors declare no conflict of interest.

References

1. Al-Haq, M.I.; Lebrasseur, E.; Tsuchiya, H.; Torii, T. Protein crystallization under an electric field. *Crystallogr. Rev.* **2007**, *13*, 29–64. [CrossRef]

2. Mirkin, N.; Moreno, A. Advances in crystal growth techniques of biological macromolecules. *J. Mex. Chem. Soc.* **2005**, *49*, 39–52.

3. Frontana-Uribe, B.A.; Moreno, A. On electrochemically assisted protein crystallization and related methods. *Cryst. Growth Des.* **2008**, *8*, 4194–4199. [CrossRef]

4. Hammadi, Z.; Veesler, S. New approaches on crystallization under electric fields. *Prog. Biophys. Mol. Biol.* **2010**, *101*, 38–44. [CrossRef] [PubMed]

5. Taleb, M.; Didierjean, C.; Jelsch, C.; Mangeot, J.P.; Capelle, B.; Aubry, A. Crystallization of proteins under an external electric field. *J. Cryst. Growth* **1999**, *200*, 575–582. [CrossRef]

6. Pérez, Y.; Eid, D.; Acosta, F.; Marín-García, L.; Jakoncic, J.; Stojanoff, V.; Frontana-Uribe, B.A.; Moreno, A. Electrochemically assisted protein crystallization of commercial cytochrome *c* without previous purification. *Cryst. Growth Des.* **2008**, *8*, 2493–2496. [CrossRef]

7. Gil-Alvaradejo, G.; Ruiz-Arellano, R.R.; Owen, C.; Rodríguez-Romero, A.; Rudiño-Piñera, E.; Antwi, M.K.; Stojanoff, V.; Moreno, A. Novel protein crystal growth electrochemical cell for applications in X-ray diffraction and atomic force microscopy. *Cryst. Growth Des.* **2011**, *11*, 3917–3922. [CrossRef]

8. Koizumi, H.; Fujiwara, K.; Uda, S. Role of the electric double layer in controlling the nucleation rate for tetragonal hen egg white lysozyme crystals by application of an external electric field. *Cryst. Growth Des.* **2010**, *10*, 2591–2595. [CrossRef]

9. Koizumi, H.; Uda, S.; Fujiwara, K.; Nozawa, J. Control of effect on the nucleation rate for hen egg white lysozyme crystals under application of an external ac electric field. *Langmuir* **2011**, *27*, 8333–8338. [CrossRef] [PubMed]

10. Pan, W.; Xu, H.; Zhang, R.; Xu, J.; Tsukamoto, K.; Han, J.; Li, A. The influence of low frequency of external electric field on nucleation enhancement of hen egg-white lysozyme (HEWL). *J. Cryst. Growth* **2015**, *428*, 35–39. [CrossRef]

11. Li, F.; Lakerveld, R. Influence of alternating electric fields on protein crystallization in microfluidic devices with patterned electrodes in a parallel-plate configuration. *Cryst. Growth Des.* **2017**, *17*, 3062–3070. [CrossRef]

12. Hou, D.; Chang, H.-C. AC field enhanced protein crystallization. *Appl. Phys. Lett.* **2008**, *92*, 223902. [CrossRef]

13. Flores-Hernandez, E.; Stojanoff, V.; Arreguin-Espinosa, R.; Moreno, A.; Sanchez-Puig, N. An electrically assisted device for protein crystallization in a vapor diffusion setup. *J. Appl. Cryst.* **2013**, *46*, 832–834. [CrossRef] [PubMed]

14. Martínez-Caballero, S.; Cuéllar-Cruz, M.; Demitri, N.; Polentarutti, M.; Rodríguez-Romero, A.; Moreno, A. Glucose isomerase polymorphs obtained using an ad hoc protein crystallization temperature device and a growth cell applying an electric field. *Cryst. Growth Des.* **2016**, *16*, 1679–1686. [CrossRef]

15. Taleb, M.; Didierjean, C.; Jelsch, C.; Mangeot, J.P.; Aubry, A. Equilibrium kinetics of lysozyme crystallization under an external electric field. *J. Cryst. Growth* **2001**, *232*, 250–255. [CrossRef]

16. Pareja-Rivera, C.; Cuéllar-Cruz, M.; Esturau-Escofet, N.; Demitri, N.; Polentarutti, M.; Stojanoff, V.; Moreno, A. Recent advances in the understanding of the influence of electric and magnetic fields on protein crystal growth. *Cryst. Growth Des.* **2017**, *17*, 135–145. [CrossRef]

17. Al-Haq, M.I.; Lebrasseur, E.; Choi, W.-K.; Tsuchiya, H.; Torii, T.; Yamazaki, H.; Shinohara, E. An apparatus for electric-field-induced protein crystallization. *J. Appl. Cryst.* **2007**, *40*, 199–201. [CrossRef]

18. Koizumi, H.; Uda, S.; Fujiwara, K.; Tachibana, M.; Kojima, K.; Nozawa, J. Crystallization of high-quality protein crystals using an external electric field. *J. Appl. Cryst.* **2015**, *48*, 1507–1513. [CrossRef]

19. Espinoza-Montero, P.J.; Moreno-Narváez, M.E.; Frontana-Uribe, B.A.; Stojanoff, V.; Moreno, A. Investigations on the use of graphite electrodes using a hull-type growth cell for electrochemically assisted protein crystallization. *Cryst. Growth Des.* **2013**, *13*, 590–598. [CrossRef] [PubMed]

20. Sazaki, G.; Moreno, A.; Nakajima, K. Novel coupling effects of the magnetic and electric fields on protein crystallization. *J. Cryst. Growth* **2004**, *262*, 499–502. [CrossRef]

21. Mirkin, N.; Frontana-Uribe, B.A.; Rodriguez-Romero, A.; Hernandez-Santoyo, A.; Moreno, A. The influence of an internal electric field upon protein crystallization using the gel-acupuncture method. *Acta Crystallogr. Sect. D Biol. Crystallogr.* **2003**, *59*, 1533–1538. [CrossRef]

22. Nieto-Mendoza, E.; Frontana-Uribe, B.A.; Sazaki, G.; Moreno, A. Investigations on electromigration phenomena for protein crystallization using crystal growth cells with multiple electrodes: Effect of the potential control. *J. Cryst. Growth* **2005**, *275*, e1437–e1446. [CrossRef]

23. Nanev, C.N.; Penkova, A. Nucleation of lysozyme crystals under external electric and ultrasonic fields. *J. Cryst. Growth* **2001**, *232*, 285–293. [CrossRef]

24. Sui, S.; Perry, S.L. Microfluidics: From crystallization to serial time-resolved crystallography. *Struct. Dyn.* **2017**, *4*, 032202. [CrossRef]

25. Ghazal, A.; Lafleur, J.P.; Mortensen, K.; Kutter, J.P.; Arleth, L.; Jensen, G.V. Recent advances in X-ray compatible microfluidics for applications in soft materials and life sciences. *Lab Chip* **2016**, *16*, 4263–4295. [CrossRef] [PubMed]

26. Sauter, C.; Dhouib, K.; Lorber, B. from macrofluidics to microfluidics for the crystallization of biological macromolecules. *Cryst. Growth Des.* **2007**, *7*, 2247–2250. [CrossRef]

27. Chavas, L.M.G.; Gumprecht, L.; Chapman, H.N. Possibilities for serial femtosecond crystallography sample delivery at future light sources. *Struct. Dyn.* **2015**, *2*, 041709. [CrossRef] [PubMed]

28. Weierstall, U. Liquid sample delivery techniques for serial femtosecond crystallography. *Philos. Trans. R. Soc. B* **2014**, *369*, 20130337. [CrossRef] [PubMed]

29. Guha, S.; Perry, S.L.; Pawate, A.S.; Kenis, P.J.A. Fabrication of X-ray compatible microfluidic platforms for protein crystallization. *Sens. Actuators B* **2012**, *174*, 1–9. [CrossRef] [PubMed]

30. Perry, S.L.; Guha, S.; Pawate, A.S.; Bhaskarla, A.; Agarwal, V.; Nair, S.K.; Kenis, P.J.A. A microfluidic approach for protein structure determination at room temperature via on-chip anomalous diffraction. *Lab Chip* **2013**, *13*, 3183–3187. [CrossRef] [PubMed]

31. Heymann, M.; Opthalage, A.; Wierman, J.L.; Akella, S.; Szebenyi, D.M.; Gruner, S.M.; Fraden, S. Room-Temperature serial crystallography using a kinetically optimized microfluidic device for protein crystallization and on-chip X-ray diffraction. *IUCrJ* **2014**, *1*, 349–360. [CrossRef] [PubMed]

32. Kisselman, G.; Qiu, W.; Romanov, V.; Thompson, C.M.; Lam, R.; Battaile, K.P.; Pai, E.F.; Chirgadze, N.Y. X-CHIP: An integrated platform for high-throughput protein crystallization and on-the-chip X-ray diffraction data collection. *Acta Crystallogr. Sect. D Biol. Crystallogr.* **2011**, *67*, 533–539. [CrossRef] [PubMed]

33. Chirgadze, N.Y.; Kisselman, G.; Qiu, W.; Romanov, V.; Thompson, C.M.; Lam, R.; Battaile, K.P.; Pai, E.F. X-CHIP: An integrated platform for high-throughput protein crystallography. In *Recent Advances in Crystallography*; Benedict, J.B., Ed.; InTech: Vienna, Austria, 2012; pp. 87–96.

34. Hunter, M.S.; Segelke, B.; Messerschmidt, M.; Williams, G.J.; Zatsepin, N.A.; Barty, A.; Benner, W.H.; Carlson, D.B.; Coleman, M.; Graf, A.; et al. Fixed-Target protein serial microcrystallography with an X-ray free electron laser. *Sci. Rep.* **2014**, *4*, 6026. [CrossRef] [PubMed]

35. Feld, G.K.; Heymann, M.; Benner, W.H.; Pardini, T.; Tsai, C.-J.; Boutet, S.; Coleman, M.A.; Hunter, M.S.; Li, X.; Messerschmidt, M.; et al. Low-Z polymer sample supports for fixed-target serial femtosecond X-ray crystallography. *J. Appl. Cryst.* **2015**, *48*, 1072–1079. [CrossRef]

36. Baxter, E.L.; Aguila, L.; Alonso-Mori, R.; Barnes, C.O.; Bonagura, C.A.; Brehmer, W.; Brunger, A.T.; Calero, G.; Caradoc-Davies, T.T.; Chatterjee, R.; et al. High-Density grids for efficient data collection from multiple crystals. *Acta Crystallogr. Sect. D Biol. Crystallogr.* **2016**, *72*, 2–11. [CrossRef] [PubMed]

37. Lyubimov, A.Y.; Murray, T.D.; Koehl, A.; Araci, I.E.; Uervirojnangkoorn, M.; Zeldin, O.B.; Cohen, A.E.; Soltis, S.M.; Baxter, E.L.; Brewster, A.S.; et al. Capture and X-ray diffraction studies of protein microcrystals in a microfluidic trap array. *Acta Crystallogr. Sect. D Biol. Crystallogr.* **2015**, *71*, 928–940. [CrossRef] [PubMed]

38. Roedig, P.; Vartiainen, I.; Duman, R.; Panneerselvam, S.; Stübe, N.; Lorbeer, O.; Warmer, M.; Sutton, G.; Stuart, D.I.; Weckert, E.; et al. A micro-patterned silicon chip as sample holder for macromolecular crystallography experiments with minimal background scattering. *Sci. Rep.* **2015**, *5*, 10451. [CrossRef] [PubMed]

39. Dhouib, K.; Malek, C.K.; Pfleging, W.; Gauthier-Manuel, B.; Duffait, R.; Thuillier, G.; Ferrigno, R.; Jacquamet, L.; Ohana, J.; Ferrer, J.-L.; et al. Microfluidic chips for the crystallization of biomacromolecules by counter-diffusion and on-chip crystal X-ray analysis. *Lab Chip* **2009**, *9*, 1412–1421. [CrossRef] [PubMed]

40. Pinker, F.; Brun, M.; Morin, P.; Deman, A.-L.; Chateaux, J.-F.; Olieric, V.; Stirnimann, C.; Lorber, B.; Terrier, N.; Ferrigno, R.; et al. ChipX: A novel microfluidic chip for counter-diffusion crystallization of biomolecules and in situ crystal analysis at room temperature. *Cryst. Growth Des.* **2013**, *13*, 3333–3340. [CrossRef]

41. Emamzadah, S.; Petty, T.J.; De Almeida, V.; Nishimura, T.; Joly, J.; Ferrer, J.-L.; Halazonetis, T.D. Cyclic olefin homopolymer-based microfluidics for protein crystallization and in situ X-ray diffraction. *Acta Crystallogr. Sect. D Biol. Crystallogr.* **2009**, *65*, 913–920. [CrossRef] [PubMed]

42. Huang, C.-Y.; Olieric, V.; Ma, P.; Howe, N.; Vogeley, L.; Liu, X.; Warshamanage, R.; Weinert, T.; Panepucci, E.; Kobilka, B.; et al. In meso in situ serial X-ray crystallography of soluble and membrane proteins at cryogenic temperatures. *Acta Crystallogr. Sect. D Biol. Crystallogr.* **2016**, *72*, 93–112. [CrossRef] [PubMed]

43. Huang, C.-Y.; Olieric, V.; Ma, P.; Panepucci, E.; Diederichs, K.; Wang, M.; Caffrey, M. In meso in situ serial X-ray crystallography of soluble and membrane proteins. *Acta Crystallogr. Sect. D Biol. Crystallogr.* **2015**, *71*, 1238–1256. [CrossRef] [PubMed]

44. Axford, D.; Aller, P.; Sanchez-Weatherby, J.; Sandy, J. Applications of thin-film sandwich crystallization platforms. *Acta Crystallogr. Sect. F Struct. Biol. Commun.* **2016**, 313–319. [CrossRef] [PubMed]

45. Sui, S.; Wang, Y.; Kolewe, K.W.; Srajer, V.; Henning, R.; Schiffman, J.D.; Dimitrakopoulos, C.; Perry, S.L. Graphene-Based microfluidics for serial crystallography. *Lab Chip* **2016**, *16*, 3082–3096. [CrossRef] [PubMed]

46. Avouris, P. Graphene: Electronic and photonic properties and devices. *Nano Lett.* **2010**, *10*, 4285–4294. [CrossRef] [PubMed]

47. Avouris, P.; Dimitrakopoulos, C. Graphene: Synthesis and applications. *Mater. Today* **2012**, *15*, 86–97. [CrossRef]

48. Wirtz, C.; Berner, N.C.; Duesberg, G.S. Large-Scale diffusion barriers from CVD grown graphene. *Adv. Mater. Interfaces* **2015**, *2*, 1500082. [CrossRef]

49. Kim, H.W.; Yoon, H.W.; Yoon, S.-M.; Yoo, B.M.; Ahn, B.K.; Cho, Y.H.; Shin, H.J.; Yang, H.; Paik, U.; Kwon, S.; et al. Selective gas transport through few-layered graphene and graphene oxide membranes. *Science* **2013**, *342*, 91–95. [CrossRef] [PubMed]

50. Li, X.; Cai, W.; An, J.; Kim, S.; Nah, J.; Yang, D.; Piner, R.; Velamakanni, A.; Jung, I.; Tutuc, E.; et al. Large-Area synthesis of high-quality and uniform graphene films on copper foils. *Science* **2009**, *324*, 1312–1314. [CrossRef] [PubMed]

51. Li, X.; Zhu, Y.; Cai, W.; Borysiak, M.; Han, B.; Chen, D.; Piner, R.D.; Colombo, L.; Ruoff, R.S. Transfer of large-area graphene films for high-performance transparent conductive electrodes. *Nano Lett.* **2009**, *9*, 4359–4363. [CrossRef] [PubMed]

52. Bird, C.L.; Kuhn, A.T. Electrochemistry of the viologens. *Chem. Soc. Rev.* **1981**, *10*, 49–82. [CrossRef]

53. Aristov, N.; Habekost, A. Electrochromism of methylviologen (paraquat). *World J. Chem. Educ.* **2015**, *3*, 82–86. [CrossRef]

54. Heyrovský, M. The electroreduction of methyl viologen. *J. Chem. Soc. Chem. Commun.* **1987**, 1856–1857. [CrossRef]

55. Chayen, N.E.; Shaw Stewart, P.D.; Blow, D.M. Microbatch crystallization under oil—A new technique allowing many small-volume crystallization trials. *J. Cryst. Growth* **1992**, *122*, 176–180. [CrossRef]

56. Schneider, C.A.; Rasband, W.S.; Eliceiri, K.W. NIH image to ImageJ: 25 years of image analysis. *Nat. Methods* **2012**, *9*, 671–675. [CrossRef] [PubMed]

crystals

MDPI

Article

Electro-Infiltration of Cytochrome C into a Porous Silicon Network, and Its Effect on Nucleation and Protein Crystallization—Studies of the Electrical Properties of Porous Silicon Layer-Protein Systems for Applications in Electron-Transfer Biomolecular Devices

Laura E. Serrano-De la Rosa [1], Abel Moreno [2] and Mauricio Pacio [1,*

[1]   Instituto de Ciencias-CIDS Benemérita Universidad Autónoma de Puebla, Ed. 130 C,
      Col. San Manuel C.P., 72570 Puebla, Mexico; lauralesd@gmail.com
[2]   Instituto de Química, Universidad Nacional Autónoma de México, Circuito Exterior C.U.,
      04510 Mexico City, Mexico; carcamo@unam.mx
*   Correspondence: mauriciopcmx@yahoo.com.mx; Tel.: +52-222-229-5500 (ext. 7737)

Received: 6 May 2017; Accepted: 23 June 2017; Published: 28 June 2017

Abstract: In this work, we report the electrical properties of cytochrome C (Cyt C) inside porous silicon (PSi). We first used two techniques of protein infiltration: classic sitting drop and electrochemical migration methods. The electrochemically assisted cell, used for the infiltration by electro-migration, improved the Cyt C nucleation and the crystallization behavior due to the PSi. We were able to carry out the crystallization thanks to the previous infiltration of proteins inside the Si pores network. We then continued the protein crystal growth through a vapor diffusion set-up. Secondly, we applied both forward and reverse bias currents only to the infiltrated Cyt C. Finally, the electrical characteristics were compared to the control (the protein molecules of which were not infiltrated) and to the samples without protein infiltration. The linker used in the sitting drop method influenced the electrical properties, which showed a modification in the current density. The simple drop method showed a current density of ~42 A/cm^2; when employing the electrochemical cell technique, the current density was ~318 A/cm^2; for the crystallized structures, it was ~0.908 A/cm^2.

Keywords: cytochrome C nucleation and crystallization; protein infiltration; porous silicon; electrical properties; silanes; electron-transfer biomolecular devices; I–V characteristics

---

1. Introduction

Porous Si (PSi) is a versatile material that has generated interest because of its compatibility with biological materials and its different applications, such as in medicine [1,2] and molecular electronics [3–5]. Cytochrome C is a protein whose solid-state electrical properties have been studied in recent years [6–8]. Thin films of cytochrome C in its solid state have conduction mechanisms both inside the molecule and at the macroscopic level in the form of a thin film [9,10]. The possibility of using these conduction mechanisms, using PSi as a base for infiltrating cytochrome C, allows the study of the thin films' electrical properties as a hybrid material—that is, an effective medium formed by PSi and cytochrome C.

Different protein immobilization methods have been reported. The most common is the drop infiltration method or covalent bond immobilization [11,12], which consists of modifying the functional group of the PSi surface by thermal oxidation and subsequently with two silane groups. This method serves as a binder to immobilize the protein, thereby forming covalent bonds of the functional groups of the PSi surface with some of the functional groups of the protein. This method also allows its position to be identified with respect to the metallic contact or the substrate, thereby allowing electrical measurements to be carried out.

In this work, we used an ad hoc method for protein infiltration, based on preceding work [13] in which the protein crystallization was induced by employing magnetic and electric fields (DC) as nucleation-inductor factors. With this method, it is not necessary to do thermal oxidation or silane impregnation of the PSi; only an anodizing current is applied that causes the protein to infiltrate into the porous silicone due to the isoelectric point of the protein and the PSi. Subsequently, we continued with the protein crystal growth through a vapor diffusion set-up. We applied both forward and reverse bias currents only to the infiltrated Cyt C. Finally, the electrical characteristics were compared to the M/Ox/Si and M/PSi/Si structures, as well as to the control (where the protein molecules were not infiltrated) in the case of the crystallized samples.

The electrical properties can be studied through the measurement of current–voltage (I–V) curves as a first approximation, to determine whether the material modifies its conductivity according to the method of infiltration. Once the protein is infiltrated by anodization, the I–V curves of the material are obtained. In comparison with the I–V curves of the immobilization, infiltration is larger by two orders of magnitude. The I–V curves of the crystallized proteins inside the PSi show a change in the order of magnitude and the current shows an asymmetric form. We then compared the level of current obtained for the crystallized protein with respect to the non-crystallized proteins. Our first approximation was based on the Schottky diode model, where the I–V curves let us determine the Schottky barrier height, considering the structure of the porous material as an effective medium.

$$Rc = \left(\frac{\partial J}{\partial V} \right)^{-1}, \text{ at } V = 0. \tag{1}$$

The porous layer is considered as in the model proposed by Vikulov et al. [14], in the equation for current density. In order to determine the height of the barrier we assume the structure to be as in the Schottky diode and use the relation of specific contact resistance (Rc), the relation for a rectifying contact in our case, according to the experimental I–V curves. This allows us to obtain the relations for Rc [15]:

$$Rc = \frac{\left(\frac{kT}{e} \right) exp \left(\frac{e\phi_B}{kT} \right)}{A^* T^2} \tag{2}$$

where k is the Boltzmann constant, T is the temperature, A* is the Richardson constant and ϕ_B is the Schottky height barrier. Following the idea of the Schottky diode model, the ideality factor (n) can be calculated using the exponential relations:

$$J = A^* T^2 \exp \left(\frac{-q\phi_B}{kT} \right) \left[e^{\left(\frac{qV_a}{nkT} \right)} - 1 \right], \tag{3}$$

where n is the ideality factor, determined by evaluating the slope in the I–V curve's exponential section.

2. Results and Discussion

2.1. SEM Images of the Porous Silicon Infiltrated Structures

2.1.1. Structures Infiltrated Using Silane-Coupling Agents

The PSi was stabilized by thermal oxidation at 1073.15 K, generating Si–O bonds. After this, the PSi/oxide layer was infiltrated employing silane-coupling agents, 3-aminopropyltriethoxysilane (APTES) and (3-Mercaptopropyl) trimethoxysilane (MPTMS), producing silane functional groups. The APTES [16] and MPTMS [17] molecule sizes are reported as 0.56 and 0.7 nm respectively. The silver (Ag) contact evaporated thickness is around 200 nm and the area covered by the circular Ag contact is 3.14×10^{-6} m^2; this area is used to determinate the maximum current density J in the electrical characterization. Figure 1 shows SEM images of the structure M/PSi-APTES-Cyt C/Si. The image in Figure 1a presents the surface of the structure that was infiltrated; the pores observed are around 10 nm in diameter. The image in Figure 1b presents the center zone in (a) that is amplified to observe the pores clearly. The image in Figure 1c shows a cross view of the structure with the metallic contact evaporated on its surface. The top of Figure 1c shows the Ag contact interface; the bottom shows the Psi–Si interface. Through this scheme of the structure we carried out the electrical characterization. The thickness of the PSi layer is around 1 μm according the scale in Figure 1c. The image in Figure 1d shows the top of the structure without metal contact. Apparently, the infiltration process distorts the first 200 nm from the top to the PSi, and this shows the penetration distance inside the porous layer.

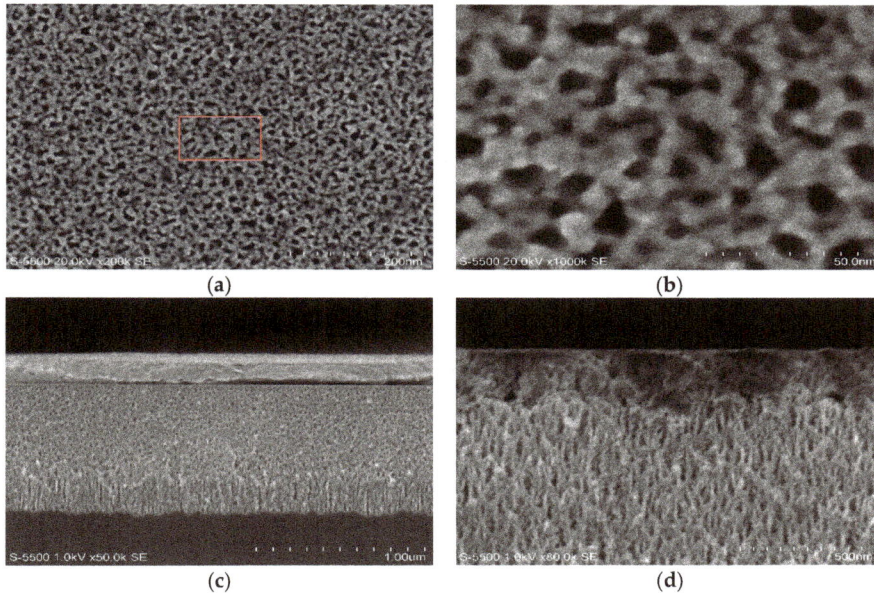

(a)

(b)

(c)

(d)

Figure 1. SEM images of the 3–aminopropyltriethoxysilane (APTES) structure: (**a**) top view of the infiltrated layer in a zone without Ag contact; (**b**) amplification of the center zone from (**a**); (**c**) transverse view of the M/PSi-APTES-Cyt/Si structure; (**d**) transverse view of the structure without metal contact.

Similar images are shown for M/PSi-MPTMS-Cyt/Si in Figure 2. The image in Figure 2a corresponds to the structure after the Cyt C infiltration. The amplification in Figure 1b shows open pores, but in Figure 2b the pores appear closed; apparently, the MPTMS agglomerates on the surface. In Figure 2c, the structure cross view shows that the Psi layer thickness is ~4 μm. In Figure 2d, a difference in tone is shown at 100 nm from the top side of the layer, which suggests that the infiltrated materials are inside this first section of the layer.

In Figure 2d, the top of layer is darker than in Figure 1d and the top layer is thinner, at ~100 nm, suggesting that a layer of MPTMS was formed over the surface. That is in agreement with the dimensions of the silane molecules. On the other hand, the cross view in both of the structures reveals that the pores are empty in the structure with APTES, while the pores in the structure with MPTMS are filled or covered.

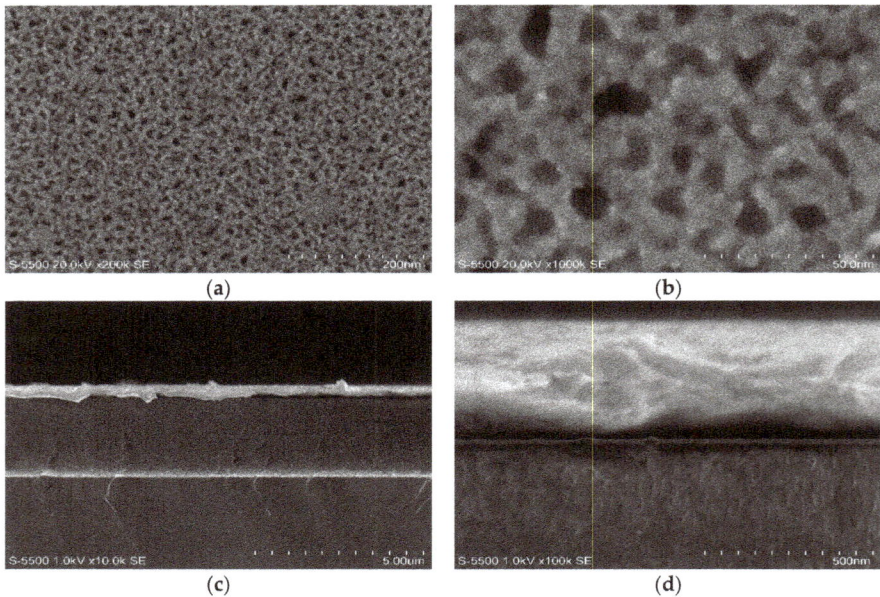

Figure 2. SEM images of the 3–mercaptopropyltrimethoxysilane (MPTMS) structure: (**a**) top view of the infiltrated layer, (**b**) amplification of the center zone from (**a**); (**c**) cross view of the structure, at the top of the image the Ag contact interface, with the Psi–Si interface highlighted by the charge accumulation; (**d**) transverse view of the structure.

2.1.2. Structures Infiltrated by the Electrochemical Migration Method

In the second infiltration method, the PSi was stabilized as described in the previous subsection, without infiltration with silane-coupling agents. The electrochemical migration method was used to infiltrate the Cyt C into the walls of the PSi. Figure 3 shows SEM images of 12 μm and 3.3 h of infiltration of the sample. In Figure 3a, the top view of the sample, pores cannot be observed as they are completely filled or covered by protein. Figure 3b shows the thickness of the PSi layer. On the top, a thicker protein layer is present, around the same thickness as the PSi layer; this explain why no pores are observed in Figure 3a.

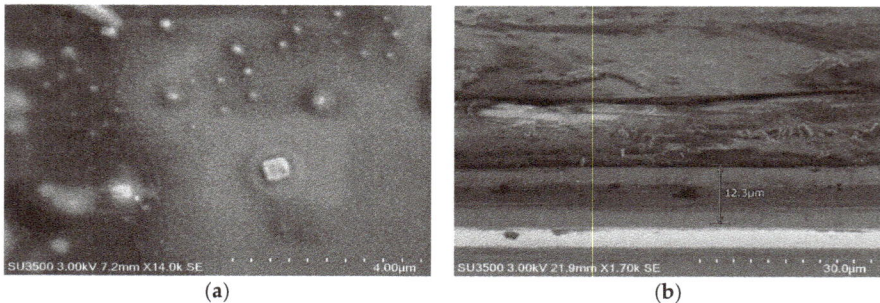

Figure 3. SEM images from the top and cross view of the sample with 12 μm and 3.3 h of infiltration: (**a**) top view of the infiltrated layer; (**b**) cross view of the structure; at the top of image, the Ag contact and the Cyt C layer formed with the electrochemical migration.

2.1.3. Structures Infiltrated with Crystallized Cytochrome C

In order to know the difference between the infiltration processes used in this work, we used a control sample, prepared using the hanging-drop vapor diffusion method (see Section 3.4), which does not present electrochemical migration in the crystallization process of Cyt C. In Figure 4, SEM images for both cases are shown. In Figure 4a, the surface of the control sample, the crystals formed have a size of ~10 μm. It should be noted that the protein in this work is not purified, so the form and size of the crystals obtained were smaller and differently shaped compared to the orthorhombic crystals reported by Mirkin et al. [18]. The image in Figure 4b corresponds to the electrochemical migration sample; in this case, the crystals are on the sample surface. They have a size of ~1 μm, and the orthorhombic shape is not well-defined. The needles in the image correspond to the surfactant residues.

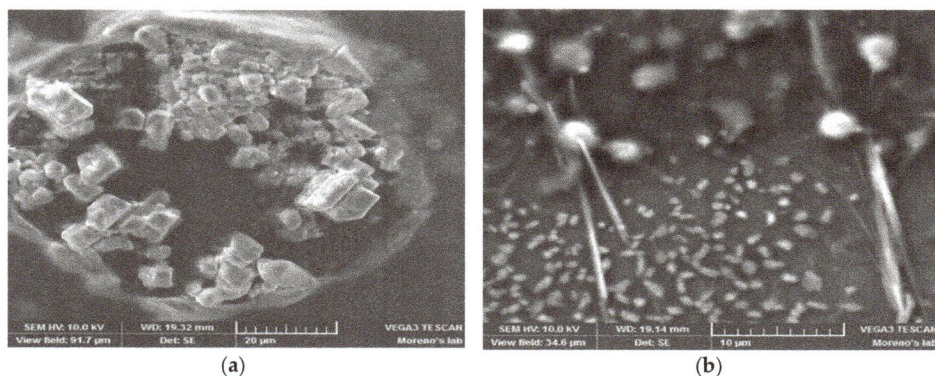

Figure 4. SEM images of the structures after the crystallization process of the infiltrated proteins: (**a**) crystals on the sample control; (**b**) micro-crystals on the surface of the PSi structure (the larger needles are the crystal salts of the precipitating agent).

The results obtained from the SEM images can be compared with optical images of crystallized Cyt C, both purified and non-purified, infiltrated into PSi. Figure 5a shows that the size and orthorhombic shape of the crystals is larger and well-defined, respectively, for the purified Cyt C; the shape corresponds to the isoform F1, in agreement with that reported for this protein elsewhere [19]. In Figure 5b, which corresponds to the unpurified Cyt C infiltrated in the PSi, the crystals are smaller, irregular and less well-defined than those for the orthorhombic-shaped crystals. The same behavior is observed in the SEM images (Figure 4).

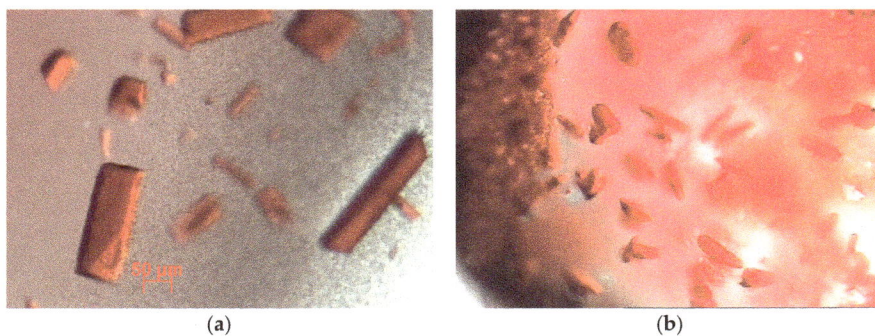

Figure 5. Optical imaging, (**a**) crystals of purified protein; (**b**) non-purified protein crystals.

The crystallization of cytochrome C was possible even with the unpurified protein; the effect of the electromagnetic field on the growth of protein crystals has already been reported [18]; however, the possibility of obtaining protein crystals with unpurified proteins opens the possibility of other applications that are more economical and simple in terms of crystallization time.

2.2. Electrical Characterization

The electrical characterization is based on I–V curve measurements. The semi-log I–V curves were measured at room temperature. To understand the conduction mechanism for the samples obtained, the log–log I–V curves for the APTES sample are shown. In the equipment used to measure the I–V curves, a compliance voltage limits the intensity of the current, so the curves show a flat region when the current reaches 0.1 A. At that point, the current density J was determined using the contact area value.

2.2.1. I–V Measurements of APTES and MPTMS Structures

Figure 6 shows the I–V curves for the M/PSi-APTES-Cyt C/Si and M/PSi-MPTMS-Cyt C/Si structures. The black line is the measurement of the M/Ox/Si structure. The graphic in Figure 6b shows the log–log I–V of the forward bias. The current from MPTMS is higher and more symmetric than that from APTES. The behavior of the MPTMS structure is closer to the M/Ox/Si.

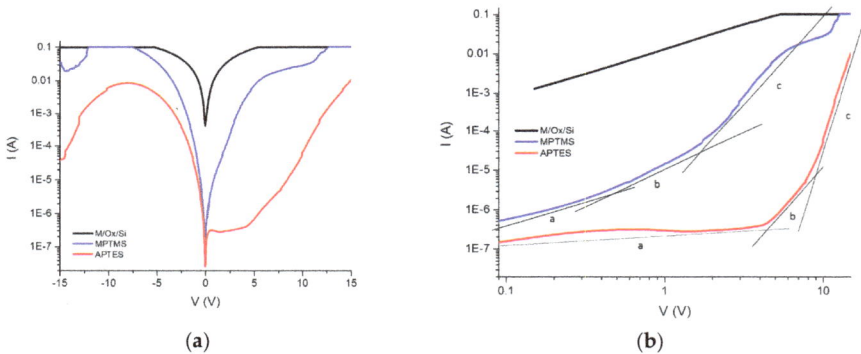

Figure 6. I–V curves for the M/PSi-APTES-Cyt C/Si and M/PSi-MPTMS-Cyt C/Si structures: (**a**) log I vs. V measures; (**b**) log–log I–V for forward bias. Sections a, b and c, indicate the conduction mechanisms described in the text.

The log–log I–V curves are clearly different for both structures. The graphic in Figure 6b considers the infiltrated PSi an effective medium, which enables us to consider three sections in the curve, related to different conduction mechanisms. Figure 6b shows the following behavior: Section (a) is associated with the deep traps at the interface of the M/PSi-silane coupling agent–Cyt C, and the dependence is linear, modifying the slope in the low-voltage region. In Section (b), the current increases exponentially, the slope is the ideality factor in the Schottky structures. In this section, the dominant conduction mechanism is recombination-tunneling [20]. In Section (c), the current follows the potential relation >>2, which indicates that the transport mechanism is dependent on temperature; this requires measurements at different temperatures [21]. Table 1 shows the slopes determined from the I–V curves for each section in Figure 6b.

Table 1. Slopes for the sections in the I–V curves for the 3-aminopropyltriethoxysilane (APTES) and 3–mercaptopropyltrimethoxysilane (MPTMS) samples.

Sample	Section a	Section b	Section c
APTES	0.18	4.62	12.92
MPTMS	1.14	2.11	4.42

2.2.2. I–V Measurements of Structures Obtained by the Electrochemical Migration Method

For infiltration by electrochemical migration, M/PSi-Cyt C/Si structures were fabricated. The infiltration times were 2.2, 3.3 and 24 h. The thicknesses are listed in Table 2. Figure 7 shows the log I vs. V curves for different times and thicknesses; curves for M/Ox/Si and M/PSi/Si structures are in both graphics for comparison. Figure 7a, the graphic for 2.2 h, shows the current intensity for layers with 10 and 3 μm thickness; the current is lower than the PSi and the shape is not symmetrical. The inset is the log–log I–V curves; two slopes are identified and the values are similar for both samples; this indicates that the electrical behavior is independent of the thickness of the PSi layer. The values for the slopes are listed in Table 3.

Table 2. Thickness and infiltration time for the electrochemical migration of the infiltrated samples.

Sample	Thichness (μm)	Infiltration Time (h)
EM1	3	2:20
EM2	10	2:20
EM3	3	3:30
EM4	6	3:30
EM5	12	3:30

Table 3. Slopes for the sections in the I–V curves for the samples in Figure 7a.

Sample	a	b
EM1	1.36	5.79
EM2	1.31	6.86

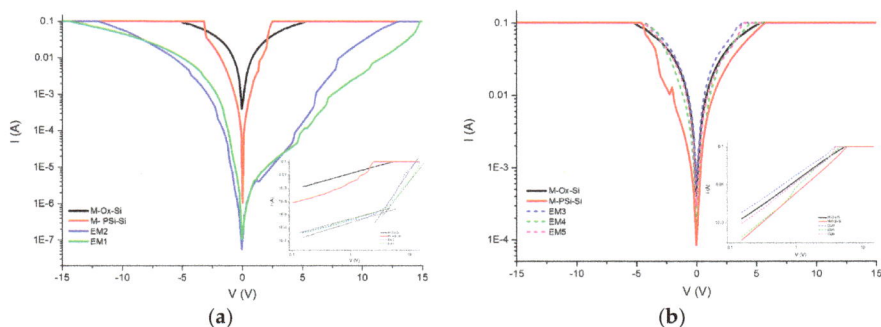

Figure 7. (a,b) I–V curves for samples with different thicknesses and infiltration times.

The slope in Section (a) for the two samples is between 1 and 2, thus the predominant conduction mechanism is recombination-tunneling. The graphic in Figure 7b shows the current intensity for samples infiltrated for 3.3 h. The slope values confirm behavior close to that of the M/Ox/Si structures, and in the case of the EM3 sample the current is above the conductive behavior of those structures.

In Table 4, the slope values are listed, including the values for the M/Ox/Si and M/PSi/Si structures. As in the case of 2.2 h samples, the behavior is independent of the thickness of the PSi layer.

Table 4. Slopes of log–log I–V curves for the samples in Figure 7b.

Sample	Slope (n)
EM3	1.21
EM4	1.32
EM5	1.60
M/Ox/Si	1.24
M/PSi/Si	1.54

The graphic in Figure 8a shows the I–V curves for the 2:20 and 3.3 h samples with 3 μm; the structure infiltrated for 3.3 h has a current higher than that of the M/Ox/Si structure. Infiltration time is a factor of the quantity of Cyt C infiltrated into the PSi for the M/PSi-Cyt C/Si structure. This agrees with the SEM image in Figure 3. In Figure 8b, the log–log I–V curves for the forward bias, obtained from the graphic in Figure 8a, allow us to compare the magnitude order of current I and to gain knowledge of the conduction mechanisms.

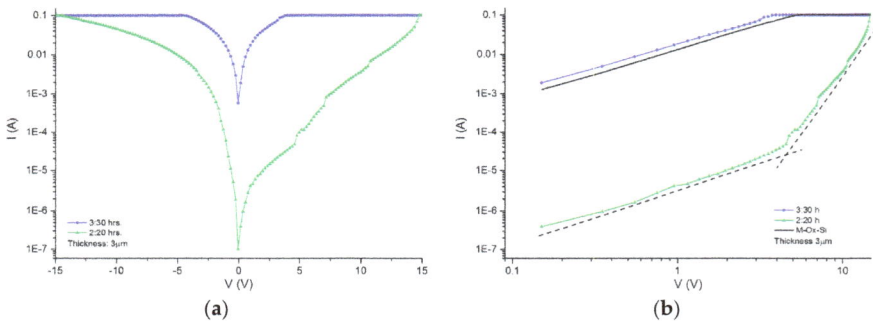

Figure 8. Comparison of I–V curves for samples with 2.2 h (a) and 3.3 h (b) infiltration times.

2.2.3. I–V Measurements of Structures after the Crystallization Process

In Figure 9, I–V curves for M/PSi-Cyt C/Si structures after the crystallization process, are shown. The infiltration time was 24 h. The black line corresponds to the control sample (M/PSi-Ox/Si structure) without Cyt C infiltration. The crystallization process reduces the current intensity and, at the same time, changes the conduction mechanisms. This behavior is close to that observed in Figure 6a. The inset graphic shows the log–log I–V curves; as in the structures in Figure 7a, only the deep trap and recombination-tunneling conduction mechanisms can be observed. The slopes from these structures are listed in Table 5.

Table 5. Slopes of log–log I–V curves for the samples infiltrated after the crystallization process.

Sample	(a)	(b)	(c)	Thickness (μm)
Control	0.88	3.72	4.42	10
XTAL 1	0.96	11.79		50
XTAL 2	1.09	6.35		5

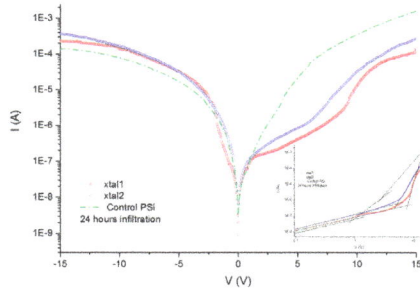

Figure 9. I–V curves of the structure samples M/PSi-Cyt C/Si after the crystallization process. The inset shows the log–log I–V curves.

2.3. Determination of Electrical Parameters

As already mentioned, the structures in this work were considered to be like a Schottky diode. The electrical parameters were determined by Equations (1)–(3). The contact resistance (Rc) values were obtained from the I–V curves; as indicated in Equation (1), this is the inverse of derivate respect to applied voltage, evaluated in V = 0. The values of Rc were used in Equation (2) in order to obtain the height barrier (ϕ_B); thus, from Equation (2), the relation for the height barrier is as follows:

$$\phi_B = \frac{kT}{e} ln \left(\frac{eA^*T^2 Rc}{kT} \right) \qquad (4)$$

Table 6 lists the electrical parameters obtained from the I–V curves; that is, Rc, ϕ_B and the maximum current density J. The value of J, as we mentioned in Section 2.3, was obtained at the point that the compliance voltage limits the measuring instrument, causing the current to become a flat line; in the case of the crystallized samples, the current was below the limit mentioned, so the value of J was determined at the maximum value of V = 15 volts.

Table 6. Electrical parameters calculated using Equations (1)–(4) and the I–V curves.

Sample	Rc (Ω)	ϕ_B (eV)	J (A/cm^2)
APTES	0.19	1.56	32.33
MPTMS	18,511.47	1.66	42.20
ME1	208.04	1.58	315.13
ME2	78.28	1.54	315.13
ME3	0.14	1.35	318.55
ME4	0.02	1.37	298.38
ME5	0.0097	1.36	310.30
XTAL1	900.65	1.58	0.408
XTAL2	2642.11	1.61	0.908

Rc: contact resistance; ϕ_B: height barrier; J: maximum current density.

The ϕ_B values listed in Table 6 do not seem to be consistent with the current density, which provides an idea about the equations used in this work; since this is a first approximation we consider that the model must be improved, considering the conduction mechanisms of both the PSi and Cyt C separately.

3. Materials and Methods

3.1. APTES and MPTMS Structures

Figure 10 shows a scheme of the process for the obtainment of the structures. In the first part, the PSi layers were obtained by electrochemical anodization. Next, these layers were placed in a furnace for the oxidation process, followed by the infiltration processes through the simple drop and electrochemical migration methods. Finally, all samples were evaporated on a metallic contact; in this case, with Ag circles. The electrical characterization was realized for all samples; the I–V measurements were taken in the transversal mode as shown in Figure 10c as published elsewhere [21].

3.1.1. PSi Preparation

The PSi layers used in this work were obtained by the electrochemical etching of p-type Si, phosphorus-doped, single-side-polished and 100-oriented silicon. Highly doped substrates (p++) with a resistivity of <0.005 Ω·cm were used. The electrolyte consisted of a mixture of aqueous 40 wt % HF (hydrofluoric acid) and absolute ethanol (99.9%) in a volumetric ratio of 1:1. An etching process was performed using a small cell made of Teflon, as described in Figure 10a. The etching process was executed to obtain layers with 60% porosity, determined by the gravimetric method [22]. The gravimetric method and profiler measurements were used to determine the growth velocity and thickness of the layers

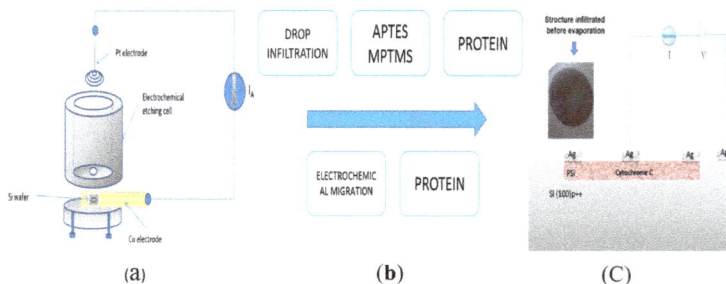

(a) (b) (C)

Figure 10. Scheme of the processes performed in order to obtain the samples: (**a**) electrochemical anodization cell; (**b**) infiltration process steps; (**c**) experimental arrangement of electrical measurement in the transversal mode. The structure is embedded in the Si wafer; the picture on the left side is from before the evaporation process.

3.1.2. Oxidation and Silane Stabilization

Before the infiltration process, the PSi must be modified due to the functional groups on the surface. The oxidation and silane stabilization processes allow the immobilization of proteins on the surface of the PSi. The thermic oxidation of PSi consists of placing the samples inside a tubular furnace (25–1200 °C) with a controller 1/16-DIN PROFILER/CONTROLLER. The process was carried out with an oxygen atmosphere: firstly, pre-oxidation at 300 °C for 10 min, followed by a temperature ramp of 100 °C/min up to 800 °C. This temperature was maintained for 3 min, at which point the furnace was turned off. The samples were removed to room temperature and stored in a desiccator.

The two silane stabilization processes (employing silane-coupling agents APTES and MPTMS) were performed: (1) the samples were dipped in 3-aminopropyltriethoxysilane (5% in toluene) for 1.5 h, then dipped in toluene and then dried with N_2 flow and stored at room temperature; (2) the samples were immersed in 3-mercaptopropyltrimethoxysilane (5% in isopropanol) as the functional reagent for 2 h. Samples were then removed and washed with isopropanol, then dried with N_2 flow and stored at room temperature until use.

3.1.3. Drop Infiltration

Once the samples were stabilized by the silane-coupling agents, a drop of Cyt C solution was deposited on the surface of the porous layer for 24 h at 277 K. After that, the remaining solution was retired and the samples were covered for 12 h at room temperature. The solution was prepared with a concentration of 5 mg/mL of cytochrome C from Sigma–Aldrich and a phosphate buffer solution of pH 7.0 and 1 M from J. T. Baker.

3.2. Electrochemical Migration Method

The electrochemical migration process consisted of a crystal growth cell connected to a DC power supply, which allows the protein to migrate inside the PSi without the silane stabilization process. The DC apparatus was a Potentiostat/Galvanostat (Vimar, Mexico) which supplied a direct current of 2 μA for 2.2, 3.3 and 24 h. The construction of the electrochemical migration cell was made following the procedure of Pareja-Rivera et al. [13], using the Si wafer as a cathode. The description of this cell is given in Appendix A.

3.3. Sample Preparation for I–V Measuurements

For the I–V measurements, all samples were evaporated on a metallic contact, using a stainless steel mask with circles of 2 mm diameter. Silver wire from Sigma–Aldrich with 99.99% purity was used at 1×10^{-6} torr. Other authors report an additional step after evaporation: their samples were subjected to heat treatment at 300 °C in order to ensure the ohmic contact. In this work, because of the protein characteristics, this heating was not possible. A Keithley 4200 A Semiconductor measuring system, using tungsten micrometric positioning tips, was used to obtain the I–V curves [21].

3.4. Crystallization Process

The crystallization of the cytochrome C used for the control sample was prepared by using the hanging-drop vapor diffusion method on EasyXtal crystal support plates (QIAGEN Lot No. 55402874). The crystallization conditions were those published by Sanishvili et al. [23], who also crystallized the oxidized form of native Cyt C. The reservoir (750 μL) contained 30% PEG-1000 in 50 mM sodium phosphate pH 7.0. Droplets of 4 μL (2 μL protein + 2 μL precipitant) containing 25 mg/mL of cytochrome C with 25% PEG-1000 both in the same buffer phosphate pH 7.0 were incubated at 277 K for 32 days. Under the conditions used by Sanishvili et al., we obtained poorly shaped aggregates. In our previous publication [18] using re-purified Cyt C, these aggregates were used as microseeds and were transferred (by means of a cat whisker) to pre-equilibrated droplets containing 22 mg/mL of protein and 25% PEG-1000 both prepared in 50 mM buffer phosphate pH 7.0. After one week, several orthorhombic crystals could be obtained

Cytochrome C was infiltrated as described in Section 3.2. It improved the Cyt C nucleation and the crystallization behavior due to the PSi. Though the microcrystals deposited on silicon were analyzed by scanning electron microscopy using a STEM Vega 3 from TESCAN (Czech Republic), they faced the risk of being destroyed when collecting images. The imaging was performed using low pressure to avoid damaging the samples.

4. Conclusions

The PSi layer was obtained by the electrochemical etching of p-type Si with 60% porosity determined by the gravimetric method. This layer was oxidized by thermal oxidation to generate Si–O bonds. These bonds were employed to pin up the silane groups obtained by the infiltration of the silane-coupling agents APTES and MPTMS. The silane groups allow the interaction with the protein and the different structures used in this work. We used two protein infiltration techniques: the sitting drop technique and electrochemical migration method. Cyt C was infiltrated into the PSi by using purified and non-purified forms. With the M/Psi-silane-coupling agent-Cyt C/Si and M/PSi-Cyt C/Si structures were obtained. The crystallization of proteins was found to be feasible on the surface of PSi. I–V curve measurements were made of these structures, which determined the Schottky barrier height. The basic equations of the Schottky diode model were used as a first approximation to determine the effect of the infiltrated protein (purified and non–purified) on conductivity into the structures. The model, according to the electrical parameters, is not the most appropriate for carrying out analysis of the structures, so the next step is to perform the I–V measurements at different temperatures and to obtain the C–V curves. However, the information obtained from this first approximation allows us to consider that a PSi layer infiltrated with a protein forms an effective media that is an alternative procedure for the building of photovoltaic devices. The electrical parameters of structures with crystallized proteins do not help to improve the transport properties within the structures, so the crystallization process can be avoided in the case of photovoltaic structures. According to the I–V curves, the most efficient method, in terms of electrical properties, is electro-migration.

Acknowledgments: The authors acknowledge the support of DGAPA-UNAM within the PAPIIT program, for the project IT200215-"Estudios biofísicos y estructurales del mecanismo de agregación/desagregación de macromoléculas biológicas. Aplicaciones a nano-biomedicina y al desarrollo de dispositivos moleculares fotovoltaicos". Laura E. Serrano-De la Rosa. Thanks to CONACYT for the PhD grant 271582.

Author Contributions: Abel Moreno and Mauricio Pacio conceived and designed the experiments; Laura Elvira Serrano De La Rosa. performed the experiments; Laura Elvira Serrano De La Rosa and Mauricio Pacio analyzed the data; Abel Moreno contributed reagents/materials/analysis tools; Laura Elvira Serrano De La Rosa and Mauricio Pacio wrote the paper and Abel Moreno wrote part of the contribution related to protein crystallization and made the final revision and corrections.

Conflicts of Interest: The authors declare no conflict of interest.

Appendix A

Electrochemical Migration Cell

The electrochemical migration cell consists of one PSi plate and the second plate was a conductive polished indium tin oxide (ITO) glass float of 3.0×2.5 cm^2, with a resistance ranging from 4 to 8 Ohms (Delta Technologies, Loveland, CO, USA). One conductive glass surface coated with indium tin oxide (ITO) and the second PSi surface were used as an electrode, placed inward, facing the structures infiltrated by electrochemical migration. The ITO and PSi structures were placed 0.5 cm from one another, in order to provide a connection that is electrically safe with the electrodes (anode/cathode) when applying a direct current (DC). The cell was prepared using a double well frame (for a vapor diffusion set up), as shown in Figure A1; the frame was made of elastic black rubber material, sealed with vacuum grease, so the rubber frame was perfectly attached to the ITO film and PSi to avoid leakage. The growth cell can be fixed using a silicone bar melting gun. The cell was constructed following the procedure of Pareja-Rivera et al. [13], but in our case using a Si wafer as a cathode.

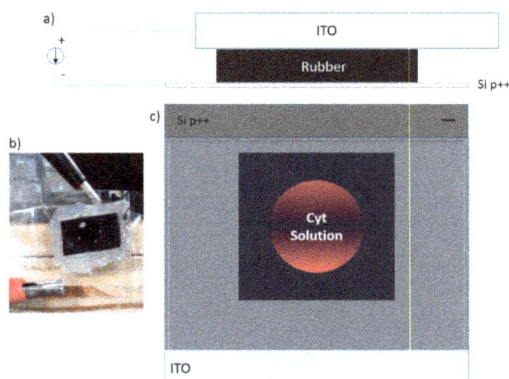

Figure A1. Electrochemical migration cell set-up scheme: (**a**) lateral view scheme of the cell, where the positive electrode is connected to the ITO, and the negative electrode is connected to the Si substrate; (**b**) image of the cell connected to the current source; (**c**) top view of the cell, where the Cyt C solution is in the center of the PSi layer area.

References

1. Ksenofontova, O.; Vasin, A.; Egorov, V.; Bobyl', A.; Soldatenkov, F.; Terukov, E.; Ulin, V.; Ulin, N.; Kiselev, O. Porous silicon and its applications in biology and medicine. *Tech. Phys.* **2014**, *59*, 66–77. [CrossRef]
2. Anglin, E.; Cheng, L.; Freeman, W.; Sailor, M. Porous silicon in drug delivery devices and materials. *Adv. Drug Deliv. Rev.* **2008**, *60*, 1266–1277. [CrossRef] [PubMed]
3. Sun, L.; Diaz-Fernandez, Y.; Gschneidtner, T.; Westerlund, F.; Lara-Avila, S.; Moth-Poulsen, K. Single–molecule electronics: From chemical design to functional devices. *Chem. Soc. Rev.* **2014**, *43*, 7378–7411. [CrossRef] [PubMed]
4. Xiang, L.; Palma, J.; Li, Y.; Mujica, V.; Ratner, M.; Tao, N. Gate–controlled conductance switching in DNA. *Nat. Commun.* **2017**, *8*, 14471. [CrossRef] [PubMed]
5. Rabinal, M. Organic molecules on silicon surface: A way to tune metal dependent Schottky barrier. *Appl. Surf. Sci.* **2016**, *382*, 41–46. [CrossRef]
6. Amdursky, N.; Ferber, D.; Bortolotti, C.; Dolgikh, D.; Chertkova, R.; Pecht, I.; Sheves, M.; Cahen, D. Solid-State Electron Transport Via Cytochrome C Depends On Electronic Coupling To Electrodes and Across The Protein. *Proc. Natl. Acad. Sci. USA* **2014**, *111*, 5556–5561. [CrossRef] [PubMed]
7. Amdursky, N.; Sepunaru, L.; Raichlin, S.; Pecht, I.; Sheves, M.; Cahen, D. Electron Transfer Proteins as Electronic Conductors: Significance Of The Metal and Its Binding Site in The Blue Cu Protein, Azurin. *Adv. Sci.* **2015**, *2*, 1400026. [CrossRef] [PubMed]
8. Ron, I.; Sepunaru, L.; Itzhakov, S.; Belenkova, T.; Friedman, N.; Pecht, I.; Sheves, M.; Cahen, D. Proteins As Electronic Materials: Electron transport through solid-state protein monolayer junctions. *J. Am. Chem. Soc.* **2010**, *132*, 4131–4140. [CrossRef] [PubMed]
9. De Groot, M.; Evers, T.; Merkx, M.; Koper, M. Electron transfer and ligand binding to cytochrome C immobilized on self-assembled monolayers. *Langmuir* **2007**, *23*, 729–736. [CrossRef] [PubMed]
10. Amdursky, N.; Pecht, I.; Sheves, M.; Cahen, D. Electron transport via cytochrome C on Si–H surfaces: Roles of Fe and Heme. *J. Am. Chem. Soc.* **2013**, *135*, 6300–6306. [CrossRef] [PubMed]
11. Kim, D.; Herr, A. Protein immobilization techniques for microfluidic assays. *Biomicrofluidics* **2013**, *7*, 041501. [CrossRef] [PubMed]
12. Márquez, J.; Cházaro-Ruiz, L.; Zimányi, L.; Palestino, G. Immobilization strategies and electrochemical evaluation of porous silicon based cytochrome C electrode. *Electrochim. Acta* **2014**, *140*, 550–556.
13. Pareja-Rivera, C.; Cuéllar-Cruz, M.; Esturau-Escofet, N.; Demitri, N.; Polentarutti, M.; Stojanoff, V.; Moreno, A. Recent Advances in the understanding of the influence of electric and magnetic fields on protein crystal growth. *Cryst. Growth Des.* **2017**, *17*, 135–145. [CrossRef]
14. Vikulov, V.; Strikha, V.; Skryshevsky, V.; Kilchitskaya, S.; Souteyrand, E.; Martin, J. Electrical features of the metal-thin porous silicon-silicon structure. *J. Phys. D* **2000**, *33*, 1957–1964. [CrossRef]
15. Neamen, D. *Semiconductor Physics and Devices*, 1st ed.; McGraw-Hill: New York, NY, USA, 2012; pp. 337–349.
16. Williams, E.; Davydov, A.; Motayed, A.; Sundaresan, S.; Bocchini, P.; Richter, L.; Stan, G.; Steffens, K.; Zangmeister, R.; Schreifels, J.; et al. Immobilization of streptavidin on 4H–Sic for biosensor development. *Appl. Surf. Sci.* **2012**, *258*, 6056–6063. [CrossRef]
17. Piwoński, I.; Grobelny, J.; Cichomski, M.; Celichowski, G.; Rogowski, J. Investigation of 3-mercaptopropyltrimethoxysilane self–assembled monolayers on Au(111) surface. *Appl. Surf. Sci.* **2005**, *242*, 147–153. [CrossRef]
18. Mirkin, N.; Jaconcic, J.; Stojanoff, V.; Moreno, A. High Resolution X-Ray Crystallographic structure of bovine heart cytochrome C and its application to the design of an electron transfer biosensor. *Proteins Struct. Funct. Bioinform.* **2007**, *70*, 83–92. [CrossRef] [PubMed]
19. Pérez, Y.; Eid, D.; Acosta, F.; Marín-García, L.; Jakoncic, J.; Stojanoff, V.; Frontana-Uribe, B.; Moreno, A. Electrochemically assisted protein crystallization of commercial cytochrome C without previous purification. *Cryst. Growth Des.* **2008**, *8*, 2493–2496. [CrossRef]
20. Çakıcı, T.; Sağlam, M.; Güzeldir, B. The comparison of electrical characteristics of Au/N-Inp/In and Au/In2s3/N-Inp/In junctions at room temperature. *Mater. Sci. Eng. B* **2015**, *193*, 61–69. [CrossRef]
21. Eltayyan, A. A new method to extract the electrical parameters from dark I-V: T experimental data of Cds/Cu(In,Ga)Se2 interface. *Int. J. Adv. Res. Phys. Sci. (IJARPS)* **2015**, *2*, 11–20.

22.   Sailor, M. *Porous Silicon in Practice*, 1st ed.; Wiley-VCH-Verl.: Weinheim, Germany, 2012; pp. 134–138.

23.   Sanishvili, R.; Margoliash, E.; Westbrook, M.; Westbrook, E.; Volz, K. Crystallization of wild-Type and mutant ferricytochromes C at low ionic strength: Seeding technique and x-ray diffraction analysis. *Acta Crystallogr. Sect. D* **1994**, *50*, 687–694. [CrossRef] [PubMed]

MDPI

St. Alban-Anlage 66

4052 Basel

Switzerland

Tel. +41 61 683 77 34

Fax +41 61 302 89 18

www.mdpi.com

Crystals Editorial Office

E-mail: crystals@mdpi.com

www.mdpi.com/journal/crystals

www.ingramcontent.com/pod-product-compliance
Lightning Source LLC
Chambersburg PA
CBHW051916210326
41597CB00033B/6167